PDA SECURITY

Incorporating Handhelds into the Enterprise

DAVID MELNICK

MARK DINMAN

ALEXANDER MURATOV

BOB ELFANBAUM

ELENA MURATOVA

DARREN RUOTOLO

DOUG STEPHAN

McGraw-Hill

New York   Chicago   San Francisco   Lisbon
London   Madrid   Mexico City   Milan   New Delhi
San Juan   Seoul   Singapore   Sydney   Toronto

The McGraw-Hill Companies

Library of Congress Cataloging-in-Publication Data

Melnick, David.
    PDA security : incorporating handhelds into the enterprise / David Melnick.
        p.   cm.
    ISBN 0-07-142490-3
        1. Computer security.   2. Portable computers—Programming.   3. Mobile
computing.   I. Title.

    QA76.9.A25M445   2003
    005.8—dc21   2003052791

1 2 3 4 5 6 7 8 9 0   DOC/DOC   0 9 8 7 6 5 4 3

ISBN 0-07-142490-3

*The sponsoring editor for this book was Judy Bass, the production supervisor was
Sherri Souffrance, and the art director for the cover was Anthony Landi. It was set in
Fairfield by MacAllister Publishing Services, LLC.*

Printed and bound by RR Donnelley.

This book is printed on recycled, acid-free paper containing a minimum of 50
percent recycled de-inked fiber.

McGraw-Hill books are available at special quantity discounts to use as premiums
and sales promotions, or for use in corporate training programs. For more information, please write to the Director of Special Sales, McGraw-Hill, 2 Penn Plaza,
New York, NY 10121-2298. Or contact your local bookstore.

CONTENTS

FOREWORD

As a child in rural Alabama, I was fascinated with the world of computers. The pristine, room-sized banks of electronic brains, whose lightning speed and flashing lights bore witness to superhuman powers, drew quite a contrast to my world where slopping hogs, chopping firewood, and hoeing okra represented the essential work of the day. I dreamt of a future dominated by computers that would, I was certain, transform the world.

Of course, my dream came true, though not along nearly as straight a path as I had predicted! And I find myself, as do many of my peers, living in a world in which our dependence on computers is pervasive. Computers (and the networks that link them) manage our livelihoods, our health, our transportation, our social lives, our governments—no areas of our lives remain untouched by information technologies. And perhaps the greatest of ironies is that I've spent an entire career working in the *information technologies* (IT) economy, most of that time managing the unintended effects of computers and networks.

Although many of the computers I rely on are relatively transparent to me, my *personal digital assistant* (PDA) is not. My PDA gives me, the most hopelessly disorganized of creatures, the luxury of a relatively structured life. My post-PDA life is one in which phone numbers and directions are at my fingertips, where pens never run out of ink, and where I never have to scrounge for a scrap of paper (which I then promptly misplace!). In order to wreak this major miracle, I've entrusted my PDA with much of the information that makes my life work. Although I have not yet taken the step of integrating my PDA with the other mainstay of my modern life, my cell phone, I am certain they will converge. Among my peers, the notion of fully participating in modern business without a PDA is novel, even inconceivable.

Despite my great affection for my PDA, I remain at heart an information security professional and, as such, am charged with worrying about the downside of technology. Thus, even as I marvel at the miraculous effect my PDA has had on my normally chaotic existence, I worry about the effects of losing control of the device itself and the information contained within it. At the same time, I see an immutable push toward enabling commercial activity and access to critical information from PDAs. This only serves to amplify my concerns.

What if someone were to take my PDA or the information it holds? In this day and age of electronic commerce, many of my business partners know me only as an email address, a user ID and password, a credit card number, an account number, or a *personal identification number* (PIN). Today a criminal, equipped with even some of the information on my PDA, can disrupt my business, harass my friends and associates, and furthermore, for all intents and purposes, *become* me, if even for only a day. Identity theft is a growing problem, and badly secured PDAs can serve as a treasure trove for those who would perpetuate this fraud.

The exposures associated with security breaches of PDAs extend well beyond the personal level. The popularity of PDAs is changing the face of corporate network security. Each connected PDA now represents a vulnerable point where attacks on Enterprise networks can be made to gain access to sensitive information. This realization has driven the fast-growing endpoint security solutions market. The proliferation of endpoints has rendered many traditional perimeter protections moot, requiring a rethinking of security architectures and protection strategies for Enterprise networks.

Therein resides the value of this book. Part of the reason it is so difficult to secure computing and networking devices is that one must first master the inner workings of the systems before applying a wide array of security assurances and features to them. Although relatively few technologists are up to this task, even fewer have the skills to write about the steps one take to protect a particular device.

In David Melnick, the reader has the perfect guide to exploring the topic of securing PDAs and other such handheld systems. I first met David several years ago when he was chief technologist for Warner Music's acclaimed e-commerce organization. I collaborated with him in designing and implementing security measures for a complex online business and network Enterprise. In my work with David, I was immediately impressed by how quickly he assimilated and then applied the security concepts and criteria I proposed. Many system integrators, even those with a strong understanding of the fundamental tenets of security, fail to understand how to apply those protections to operational systems. David had no such problem, and I was impressed by how easily he generalized mechanisms designed for one platform environment to others without losing their protective qualities. This is the skill set needed to morph existing security principles to fit PDAs and other small computer systems.

This book represents a rare opportunity for the interested user to learn the full feature set of a PDA, while also understanding how to manage those features so that they don't come back to haunt him or her later. For the Enterprise security manager, it offers an opportunity to characterize and mitigate the risks associated with the connection of PDAs to the Enterprise network. And for the PDA designer, the book represents an opportunity to see which security features might be added to new devices in order to make them safer receptacles for critical information.

Computing and networking technologies offer us the capability to do amazing things. This capability can be as breathtaking as viewing galaxies unimpeded by atmospheric diffraction or as commonplace as paying for everyday purchases using a piece of plastic with a magnetic strip affixed to it. My PDA offers me and many other users the capability to do something in its own way equally as amazing: to live a saner, better organized life. These amazing powers, well secured, can be ours for years to come.

Rebecca Gurley Bace
Scotts Valley, California
April 2003

ACKNOWLEDGMENTS

This book reflects a team achievement for a group of highly skilled professionals in fields from encryption to intrusion detection to policy management, all supporting my efforts to weave together a comprehensive discussion of what it means for organizations to address handheld computing security. I want to thank the team represented here for their contribution and support in creating this book.

Most importantly, I want thank my wife Kerri whose quiet strength inspires me to be a better man; and without whom this book would never have seen the light of day. Also, my daughters Skyler and Devon and their effortless smiles as always, continued to provide great inspiration to me through my many nights on this project.

Special thanks also to Steve Elfanbaum whose friendship ultimately in an unusual series of events also made this possible; and who over the years has provided me with great conversation and inspiration.

Finally, I want to acknowledge Becky Bace who was really the first person to introduce me to the formal security discipline. Through her I was finally able to place all risk management related experience I had developed into a framework and begin down a clear road toward the security profession.

David Melnick

I would like to acknowledge the efforts of the Asynchrony PDA Defense development team, particularly Ron Foley and Alexander Muratov. Without Ron's original idea and Alex's efforts to turn that idea into a working product, PDA Defense and my involvement in the world of Personal Digital Assistant devices would never have happened. Most importantly, I would like to acknowledge and thank my wife, Michele, and my two children, Jonathan and Bryan, for their patience and support over the last three years since making the career change and joining Asynchrony.

Mark Dinman

Thanks to my sweet wife Elena for her support, patience, and help in creating this book. Her energy and ideas support everything I undertake. Additionally, personal thanks to Nick Trokhan for his participation in PDA Defense development, and thanks to J. Mark Lambright who originally introduced me to PDA development.

Alexander Muratov

Thanks to my parents for their love and support. And thanks to my small son Michael for all the pleasure which he gives to me. Finally thanks to the entire PDA Defense development team.

Elena Muratova

Thanks to my Mom and Dad for all of their help and to my brother, The Chemical Super Freak, for challenging me by already publishing.

Darren Ruotolo

I would like to acknowledge Steve Elfanbaum, Nate McKie and David Elfanbaum, who with me, co-founded the Asynchrony Software community (www.asynchrony.com). PDA Defense is one of the highlights of the Asynchrony community, that has over 1,000 software products under development and more than 30,000 participants from 160 countries. Ron Foley and Alex Muratov (who is a co-author of this book) were the initial key parties in the conception and development of PDA Defense. The initial vision of Ron and Alex, combined with the extensive input and direction from both members of the Asynchrony community, as well as the many customers of PDA Defense, allowed us to understand the requirements of the market and build a product that proactively addresses mobile device security.

I would also like to thank my wife Valerie, and daughters Jennifer, Madelyn, and Alexa, who have been very patient as their middle-aged husband/father pursued the entrepreneurial dream. It is truly rewarding and enjoyable to go after our vision with my two brothers, but as with many things, there is always a cost to the family in supporting such dreams. I am truly grateful for their patience as we continue to grow our company, cherish our successes, and learn the hard lessons over and over again.

I would also very much like to thank all of the employees of Asynchrony Software and Asynchrony Solutions (which includes the PDA Defense division). Without the hard work and dedication of these people, we would not be able to accomplish the amazing technical solutions they deliver to our customers every day.

Bob Elfanbaum

INTRODUCTION

When we started this book with Judy Bass from McGraw-Hill, all of us were aware of the lack of good information available on PDA security. A number of wireless-security-related books are available that peripherally address PDAs but always from a network perspective and usually with a focus on transport-level security. The primary challenge organizations are facing today does not have to do with transport-level security. The focus needs to be on device security for the rapidly evolving PDA market and how to manage standards, policy, and enforcement of risk reduction.

I spend time every week with talented security, *information technology* (IT), and business professionals who are struggling with how to handle the area of handheld computing. Every organization has accepted the proliferation of these devices within their organization, except, perhaps government entities, which are the only ones with the moxy to try to ban them completely. However, increasingly difficult challenges have resulted from organizations purchasing larger and larger numbers of PDAs for an increasingly wide variety of internal business uses.

What to Expect in This Book

This book provides a comprehensive primer for organizations struggling with how to address what they have accepted as the inevitable age of handheld computing within their department, organization, and industry. IT departments, security professionals, and business executives all need to understand how to respond to the adoption of these devices as a mission-critical part of their organizations' computing needs.

This book ambitiously attempts to provide a comprehensive working tool to empower a professional to come up to speed and make informed decisions regarding PDAs. Many of the professionals who will read this book are responsible for PDA deployment, while they admittedly have had little to no preparation to make informed decisions about the challenges they face. It's not their fault! This area is evolving rapidly and no one has it all figured out yet. Experience in disciplines such as computing asset management, security (including authentication and encryption), and a general interest in handheld devices will all help, but no one can be fully prepared.

To provide the immersion experience in the range of issues a professional should tackle for managing handheld strategy or just a PDA purchase within an organization, we provide the following agenda both for this book and as a suggested set of topics for professionals organizing efforts to address PDAs within their organization:

- Overview of the industry players and emerging issues.
- Framework for assessing the risk PDAs present to your organization.
- Discussion about developing a measured policy response.
- Case example of what to expect in the tools and how to enforce security on handheld devices (on-device security).
- Technical immersion in key areas and relevant device operating systems.

Outline of How to Use This Book

The book has been divided into four sections. Each addresses a different aspect of PDA security and focuses on a different discipline in order to provide a general orientation for those responsible for PDA-related strategy and execution within their organization. The following outlines the intended use and general makeup of each section and the included chapters.

Section 1: Let's Get Oriented (Chapters 1, 2, and 3)   The first section starts at a pretty high level, addressing how the evolution of the PDA market and IT security have led to the issue of PDA security. It's a pretty quick read through Chapter 1, *"Incorporating Handhelds into the Enterprise: The Road to PDA Security,"* and Chapter 2, *"PDAs in the Enterprise: Security Becomes a Focus,"* as we set the stage for the current business challenges around PDA security. Chapter 3, *"The Power Resource Guide to Understanding Where Security Must Be Achieved,"* finishes the section with a reference guide to the operating systems, software categories, and hardware devices that make up this rapidly evolving market segment. This section as a whole serves as a general orientation to prepare you for the increasingly targeted discussions of Sections 2 and 3.

Section 2: Risk Assessment, Policy Making, and Enforcement (Chapters 4, 5, and 6)   Section 2 provides an almost standalone read for a relatively experienced IT, security, or business professional trying to plan out how to implement PDAs within their organization. Each chapter takes you through the risk assessment, policy making, and execution stages of rolling out PDA security.

Chapter 4, *"When a Handheld Becomes IT's Problem: Assessing Your Corporate Risk Profile,"* starts with a comprehensive look at assessing risk within an Enterprise organization. This chapter provides a framework to address PDA security within the context of the broader organization's security requirements. This area will be a comfortable and familiar discussion for a security professional, with just the PDA-specific twists necessary to adapt the PDA strategy into the overall security approach. For those readers less familiar with security as a discipline, this hard-working chapter lays the groundwork and provides an adequate overview addressing not only a PDA rollout and security, but also the broader security framework within which this topic should be understood.

Chapter 5, *"The Components of a Measured IT Response,"* builds on the risk assessment work of Chapter 4 and provides a practical framework for creating PDA security requirements.

However, the chapter takes clever dives into slightly technical and unique PDA issues that must be incorporated into the policy requirements of handheld computing. This chapter provides a great blend of technical and business issues around policy management and single-handedly arms an experienced professional to incorporate PDAs into the broader organizational security approach.

Chapter 6, *"The How-To Guide: Asynchrony's PDA Defense, Palm's Choice for the Enterprise,"* leverages a case-study framework focused on the deployment of an end-device security solution for a multidevice, multiplatform organization. By leveraging PDA defense as the PDA security product, the chapter walks through the policy configuration around authentication, encryption, and bit-wiping, as well as other areas to illustrate a comprehensive approach to implementing a security solution to a PDA community.

Section 3: Getting Technical (Chapter 7 and 8) In an extremely technical discussion, we provide the tenacious professional with an opportunity to get an in-depth understanding of the technology and issues around achieving security on the PDA. These chapters explore a high-level review of the core technology involved in delivering PDA security, including encryption, authentication, and exploiting vulnerabilities, to name a few. These discussions are approachable by a business user, but they also offer a deeper technical opportunity for those with the math skills, white-hat hacking interest, or just plain curiosity. If you ever wondered how encryption works at a bit level or wanted to know the mathematical model for calculating how many years it would take to crack an eight-character password, this section is for you.

Section 4: Wrap Up with a Little Discussion of the Future (Chapters 9 and 10) By this point, we shift gears again, attempting to bring our efforts together to provide a glimpse of what to expect in the months or years ahead. The handheld area continues to rapidly evolve, and a good policy is based on the crystal ball a professional brings to his or her planning session.

A Final Thought

The field you are exploring is more dynamic and interesting than any since the 1996-1997 Internet business explosion. Handheld computing may well loom larger than the desktop world in the years to come, so I hope you share my excitement and enthusiasm for exploring the field.

My final comment before we dive in has more of a cautionary tone. PDA security has become an Achilles' heel within an Enterprise's overall security strategy. We are now part of an effort to first incorporate PDA security into our overall security strategy efforts and then to actively address our nation's need to protect our national computing infrastructure. We are all mostly bit players in a grand effort to redress years of such a rapid integration of technology that control within our computing environments in most cases has become a reluctant oversight. This effort is comparable to the strong documentation that most technical professionals always plan to keep up-to-date.

As we role out PDAs in a post 9/11 world, we are fortunate to finally have greater support for addressing asset management and protection. Further, this heightened awareness and the still-emerging hardware and software tools have focused unprecedented bottom-up attention on achieving enforceable security policies within the handheld computing industry. I am cautiously optimistic that these trends will enable us to establish a handheld computing infrastructure that safeguards rather than threatens our national infrastructure.

INTRODUCTION TO PDA SECURITY IN THE ENTERPRISE

The first section of this book, including Chapters 1 and 2, will provide a broad introduction to the topic of PDA Security within the context of the Enterprise organization. While PDAs have become popular as *personal information managers* (PIM) with a mass marked consumer oriented industry, the advent of handheld computing platforms as a viable Enterprise computing platform has followed more recently.

We will use the introductory section to define our terms and orient you to both the handheld computing movement within the Enterprise and what it means for IT groups challenged with managing, securing, and broadly setting computing policy for these devices.

CHAPTER ONE

INCORPORATING HANDHELDS INTO THE ENTERPRISE

THE ROAD TO PDA SECURITY

During the course of this book, we will focus mostly on the professional tasked with *personal digital assistant* (PDA) security issues. Such a responsibility includes assessing an organization's need for security measures, the appropriate level of standards and policies for achieving security goals, and finally, instructions on how to achieve a manageable and enforceable rollout of PDA security. This chapter begins our discussion by examining the development of the PDA within the Enterprise. This development has led *information technology* (IT) organizations to focus not only on management and support, but also on security for corporate-sponsored PDA deployments.

WHAT IS A HANDHELD? LET'S DEFINE OUR TERMS

Handheld computing has advanced rapidly from clunky techie toys to early consumer success in personal information management to the diverse array of devices available today. Although consumer PDAs have become well-known mass market tools, the Enterprise has only recently started taking an interest in the large-scale, organized purchasing of handheld computing devices. The reasons stem from a number of factors, including the increasing computing power, memory, and wireless capabilities that have in turn resulted in an ever-increasing range of business applications.

In this book, we will tend to use the terms handheld computing and PDA somewhat interchangeably. PDAs are best understood as a subset of the handheld computing area, but because the devices easily labeled PDAs offer a variety of uses (leading to corporate purchasing), for the purposes of this discussion the distinctions won't be important. However, with the relative newness of the handheld device and the continued, rapid expansion of the hardware and software available, it will be valuable to spend a bit more time defining what the PDA represents.

PDAs: The Basic Personal Information Manager

Although a wide range of PDA devices were developed and marketed throughout the 1990s, they emerged to contain a rather narrow set of form factor (user interface elements such as screen shape and size, number of and placement of buttons, shape, and size of device, etc.) and software application elements. These components together formed what generally is referred to as a *personal information manager* (PIM). Although we will discuss this further in Chapter 3, *"The Power Resource Guide to Understanding Where Security Must Be Achieved,"* the PIM basically has a screen for display, a keyboard or area for Stylus-based writing, and a specific set of core applications. The basic applications were largely modeled on day-planning systems such as Franklin Planner, which primarily included a calendar, contacts manager, task planner, and an area for notes. Much of what the PDA has become in its many varied implementations stems from and still contains these basic elements.

Handhelds With an Increasing Range of Applications

With the success of the Palm platform in the mid-1990s, as will be further discussed in Chapter 3, development communities began to look at ways of leveraging the handheld computing device, which most consumers were relying on for PIM. With the mass market adoption of handheld computing devices that offered increasingly powerful *central processing units* (CPUs) and expanding memory capacity, developers began creating a wide range of added utilities to extend the value of these devices.

PDAs began to include a wide range of business productivity applications. The form factor had some limitations as a data entry device but was effective for read-only applications with limited write requirements. Information repositories of all kinds began to be incorporated into Palm applications, as it was the dominant platform, and various tools emerged for viewing and editing Microsoft Office documents. With the current list of Palm applications numbering well over 10,000, even the directory that Chapter 3 provides is a limited set of examples and categories of applications.

Compaq, now Hewlett-Packard, emerged with the first breakthrough device for Microsoft's Windows CE 3.0 platform: the iPAQ. It included a generally higher-end configuration of CPU and memory. For the first time, hardware and software developers already comfortable working with Microsoft operating systems began to look at ambitious applications for the business marketplace.

Wireless Capabilities

On a different note, the convergence of the Internet and wireless technologies became an investment frenzy for the venture capital community. As you may have experienced, you couldn't work with a *venture capitalist* (VC) that wasn't asking you about your wireless initiatives. You could be manu-facturing underwear and they would wonder how you were going to leverage wireless web capabilities!

A benefit of this irrational run at what was believed to be the "next big thing" by the investor community was an accelerated adoption of wireless technologies by manufacturers and soft-ware developers. Web browsers became core components of every PDA, regardless of whether the PDA had wireless capa-bilities built into it or not. Hardware manufacturers ensured their devices could support a wide range of *radio frequency* (RF) standards. This trend has continued to the present day with cur-rent devices' emerging standards. These include *Infrared* (IR), Bluetooth (short-range *local area network* [LAN] RF), 802.11b (wider-range LAN RF), and cellular capabilities (*Code Division Multiple Access* [CDMA] or *Global System for Mobile Commu-nications* [GSM]/*General Packet Radio Service* [GPRS]). All of these can be built into a single device.

While PDA devices were incorporating increasing numbers of client-based methods for wireless connections, organizations also began adopting, in a largely disorganized fashion, the use of 802.11x wireless network standards in access points. This facilitated the connection of various types of client devices.

Amid all the experimentation and misdirected efforts, some truly inspired implementations emerged. The BlackBerry device from *Research in Motion* (RIM) leveraged the low-bandwidth but highly reliable pager networks for the transmis-sion of email to always-on devices that initially resembled a pager more than a PDA. The BlackBerry implementation with its thumb-based keyboard changed the way other PDA manu-facturers thought about wireless and introduced perhaps the first killer application for wireless PDAs.

ROLE OF PDAs IN THE ENTERPRISE

The expanding role of PDAs in the Enterprise has been driven by their utility as a *Personal Information Manager* (PIM) as well as the success of Palm's initial breakthrough form factor and general PIM applications initially targeting consumers. Enterprise organizations have effectively exerted no leadership in guiding the direction or functionality of these devices until recently. Consumers have brought these productivity tools into organizations for one of two primary reasons: to stay connected through SmartPhone, two-way paging, and email, or to manage their calendar, contacts, and tasks.

Proliferation of PDAs

Many Enterprise organizations went through a phase in the 1980s when they purchased Franklin planners for employees and trained them on how to gain productivity through personal information management. During this time, most organizations informally supported the adoption of PDAs as a productivity tool by employees. Due to their relatively high price and initial image as a technology gadget, most organizations did not start purchasing them for employees. However, because organizations are ultimately a collection of individuals, the mass market popularity of PDAs inevitably created a high percentage of them within organizations.

PIM: From Franklin Planner to the PDA   Initially, the use of PDAs within organizations mainly focused around personal information management. Employees who were used to carrying their day planners to meetings began noticing increasing numbers of PDAs in the meeting room. The PDA became a tool for managing contacts, calendars, notes, and tasks, and although employees might periodically beam each other a contact, until the last year or two they rarely strayed beyond personal information management in their primary use.

Staying Connected: Pagers, Email, and SmartPhones   In parallel with the rapid onset of PDAs for PIM across organizations, a new set of devices with a different role also emerged for the purpose of exchanging information. Developed more out of the pager/two-way pager world than from the now mass market PDA space, devices for exchanging email messages and integrating limited PIM-type functions into a primarily pager-like device began to appear in large numbers. These devices, most noticeably symbolized by the BlackBerry from RIM and in increasing varieties of phones, represent a converging area of what generally still bears the label of PDA.

Enterprises Get Organized Around PDAs

Although the movement bringing PDAs into the workplace has been driven by individuals with their personally purchased devices, some organizations have begun to reimburse and directly purchase devices for several reasons. More recently, these purchases have started to evolve from small departmental purchases to more carefully evaluated corporate-wide purchasing based on standards and policies, and driven by a specific application or use.

Expensive Brochureware   Many early purchases of PDAs were made to provide a portable medium for delivering electronic brochures. Lexus, for example, acquired 3,000 devices for use by salespeople at dealerships. The idea was that they could use these devices in conjunction with their computers to download the latest vehicle information off the publicly available web site. The salesperson would start with their desktop computer connected to the Internet in order to access the public Lexus web site and download recent vehicle information to their computer. From there they could transfer the information to their PDA in order to have the information available to them while they are working with customers on the showroom floor.

Brochureware applications vary in purpose but tend to use the PDA largely as a read-only device for storing relatively nonvolatile information that is required as a resource. A popular and useful example includes the standard use of Epocrates by medical professionals on their Palm devices. This large database of information on medications and interactions has proven to be an early killer application within the medical profession, even though it largely provides access to a relatively nonvolatile repository of information. The key in this example, as in many others, is the mobility with which the information can be accessed.

Anywhere Email   The PDA as a communicator has become another key role, driving organized purchasing within companies. BlackBerry and their Enterprise solution, which includes a server component along with email capabilities, form a great example of a solution that requires corporate sponsorship and involvement. The solution demands the support of IT in both allowing the traffic through the corporate firewall to communicate with the RIM server solution, and managing the server infrastructure that synchronizes BlackBerry activity with the corporate email infrastructure. The result is email synchronization that delivers real value to executives and mobile workers within an organization.

Killer Applications: Sales Force Automation and EIS   The real driver of corporate-sponsored purchasing and PDA support will come from the identification of truly mission-critical business applications operating on the PDA. Although the previous examples may be deemed mission-critical by some organizations, generally most have yet to find uses for PDAs that are classified as mission-critical.

Many Enterprise organizations are looking at handheld devices and are grappling with how to set standards and determine rules for purchasing. Without a killer application driving the process, however, their efforts won't move beyond strategy documents. In organizations with strong corporate purchasing controls, an IT group can implement standards to limit the range of devices allowed for departmental purchases and perhaps require certain software components. The downside is that in the end this forms only a preliminary response to the coming handheld era.

However, it is the killer application that results in an approved budget for devices. This killer application motivates IT staff to establish a support and management approach, and leads the security group to set standards for device controls. It results in asset management software being selected and implemented to support the deployment. The killer application also results in the selection and purchase of PDAs in large numbers. Moving forward, killer applications will fuel corporate PDA buying within the industry.

Siebel and their mobility solution for their *customer relationship management* (CRM) software platform have been a driving killer application for organizations I have dealt with. CRM software *sales force automation* (SFA) applications provide a good example, because sales as a function appears to be leading the way in driving large corporate buys of PDAs. Although efforts may start with the need to deploy a killer application, along for the ride comes the purchase of the device and support software, as well as the need to establish policies, management controls, and helpdesk support. In short, from the first killer application, the IT group puts the infrastructure in place to support PDAs as a corporate asset.

Use of PDAs by Industry and Functional Area

Although every individual can benefit from the PIM capabilities of a PDA, PIM alone has not motivated IT staff to purchase PDAs for their employees. Something more is required to generally result in a corporate-sponsored purchase. In addition, the introduction of corporate-sponsored PDAs has been uneven by functional area and industry.

Several factors, including the mobility of the employees, the importance of on-demand access to information, and the possibility of replacing a laptop with a handheld, are all examples of factors that have contributed to certain organizations or groups within organizations being early adopters of PDA technology.

Use of PDAs in Department or Functional Areas   Department or functional variance in the corporate adoption of PDAs has been affected most dramatically by the mobility of workers and the associated reliance on laptops for access to information while away from the office. Both mobile executives and sales personnel demonstrate the two most frequent examples of the corporate purchasing of PDAs.

Sales Force and Mobile Workers   Sales force workers often have laptops as well as high mobility requirements. The salesperson also often strives for a competitive advantage and is an early adopter of technologies if they provide a true advantage in the selling process. Leveraging the technology of a PDA to be slightly better and faster resonates well with the general salesperson.

A stated long-term objective of a number of business leaders I have spoken with on this topic is for the ultimate replacement of laptops with the relatively less expensive and certainly more mobile PDA for sales professionals. Although I have yet to be involved in a rollout where this was an achieved objective, each progressive opportunity and device type appears to bring organizations closer to achieving this goal.

Executives Stay Connected   The BlackBerry example previously described for enabling remote email, especially when combined with a cellular phone, provides another form of functionality that has driven corporate purchasing. Most of these examples I have been directly involved in were pilot programs only for limited senior executive perk programs; however, I am in more conversations with organizations considering converged device rollouts in the thousands.

The converged device moves beyond the often illustrated BlackBerry example to the increasing number of converged devices providing PDA and phone integration. Although the best implementations I have worked with so far have been based on the Palm OS, with Microsoft's incredible commitment to this area, I fully expect Pocket PC-based devices will be ready for major rollouts in the near future.

Kyocera and its 6000 and 7000 series SmartPhones have been breakthrough devices for bringing PDA and phone capabilities into a single device that sales organizations are beginning to roll out. Samsung has also emerged as a pioneer in converged devices, aggressively moving on both the Palm and Microsoft front.

Industry-Specific Adoption   Industry differences with respect to the adoption of PDAs have been tied to the nature of the employees' work more than any other single factor in my experience. For example, organizations with extremely large, well-educated sales forces distributed across a wide geographic area are an excellent target. My experience has been tied to the use of PDA security tools for the government and military, as well as in the healthcare field.

WHO'S MANAGING AND SUPPORTING THESE THINGS?

As handheld computing becomes sponsored by corporations, much like the early stages of the personal computer before it, the adoption of this technology will not immediately come through the IT organization. Many of my early-stage customers are department leaders, such as a VP of sales. I say early-stage because as corporate sales of these devices expand, more rapidly than the case with PCs, IT organizations assert their roles. Many of the IT organizations I have helped deal with the expansion of PDAs in their organization struggle with how to provide the support and management that departments have requested.

In the beginning, PDA support meant going to your cubicle buddy who's had a PDA a little longer than you and asking him or her how to solve a problem. Over time, someone emerges as the geekiest techno-type of the company and becomes the technical guru, dolling out advice and support as requested. However, these strategies for providing technical support in a large organization break down when applications move beyond basic PIM and become mission critical. At that point the support function becomes sufficiently complicated that it requires more organized and structured support.

Because organizations are required to distribute third-party or custom software and have specific requirements around security, even our techno-gadgethead, who still may think he has all the answers, inevitably lacks the experience and knowledge to address the challenges that employees will face. Requirements to reset passwords for locked-out devices and to ensure that the most current version of software is provided tend to escalate when an IT organization is unprepared to meet them.

Professionalizing Management and Administration

As IT groups grapple with managing and supporting larger groups of often diverse PDAs across an organization, they quickly start asking all the right questions. With previous experience in the more mature desktop field, they must now seek help for managing the distribution of software, auditing the status of devices, validating versioning, enforcing software licensing requirements, locking out access, unlocking access, and basically providing management and support.

Unfortunately, this capability, with little exception, does not come out of the box with most devices. Although RIM's focus on server solutions makes it a close contender and a popular platform with IT managers, generally third-party management software must be acquired, often with price tags approaching $100 per device. Although a range of choices exist, which I will touch on shortly and explore further in Chapter 3, HP Open-Views, IBM Tivolis, and others are just starting to move into this area. IT managers are thus forced to evaluate and adopt new vendor relationships and infrastructures to address this area.

Industry Support for Management and Support

One of the challenges for Enterprise organizations trying to implement a comprehensive approach to PDA management is the limited support available from hardware and software providers. The emerging industry segments that have aligned around the PDA market have primarily focused on the challenges as an end-consumer opportunity. As Microsoft perhaps best illustrated as it transitioned in the 1990s (and still attempts to transition in some ways), Enterprise organizations have different needs and requirements from vendors than a consumer does.

Even though key players like Palm attempted to organize Enterprise-focused efforts over the last year or two, economic challenges coupled with the emerging or limited amount of corporate driven PDA purchases has led many to drop or marginalize such efforts. This happened despite the ever impending rapid adoption that many forecast to take place across corporate America. Palm in its layoffs during February and March of 2003 substantially reduced its commitment to direct domestic Enterprise sales with a refocus on *Value Added Resellers* (VARs) as their primary method or channel for selling to corporate markets.

Total Cost of Ownership

As IT organizations have begun to look at the challenge of deploying, managing, and supporting PDAs in their organization, *total cost of ownership* (TCO) numbers have begun to emerge helping us understand the true effort around addressing PDAs in the Enterprise. As most IT people would expect, TCO has a little to do with the initial sticker price of the device. PDAs are continuing to come down in initial price from the average $500 of most Enterprise-focused devices to the $99 lower-end devices entering the market. IT managers find an often referred-to $3,000 TCO number used as a guideline for including the hardware, software, maintenance, and support for the first year of a new device. I don't know what the average depreciation period should be on these capital purchases, but I would be surprised if their use-life even makes it to the three-year mark that has become a common span in the PC world.

The price components unfortunately include what often appear to be hidden costs overlooked by the project manger during a large rollout. These costs are only discovered well after budgets are final but well before the rollout is complete. Hardware costs, previously around $500, have started to drop and probably will come in at \$300 to \$500 for most projects. In addition, the killer application may add another \$100 to \$300. So with a \$400 to \$800 fully loaded cost, from where does the \$3,000 number come?

Well, this is where the IT staff needs to address system management and security. System Management software will add an average of \$100 per device with another \$20 to \$60 for security which is often overlooked during initial budgeting processes. From here, we spend most of the rest of our money dealing with maintenance, support, and upgrade issues. However, I would be remiss if I did not add in the allocation for lost, broken, and stolen devices that has a larger implication here than in most other computing areas, due to the nature of the device's form factor and use.

ENTERPRISE MANAGEMENT FOCUSES ON SECURITY

As Enterprise organizations take over the challenge of managing and supporting PDA deployments, security emerges as a core component of the conversation. When speaking with Enterprise customers, I commonly hear the terms "system" or "asset management" used in the same breath as security. Many of my relationships with asset management software providers have developed directly from their interest in management and security being part of a well-integrated single solution.

The level of an organization's focus on security depends on a number of factors that we will explore throughout this book. Security at a minimum has become a checkbox on any corporate plan to roll out PDA devices. This book will help you understand how to evaluate security issues for the PDAs within your organization and how to chart your course through the process of establishing and enforcing policy around your PDAs.

CHAPTER TWO

PDAs IN THE ENTERPRISE

SECURITY BECOMES A FOCUS

With enterprise *information technology* (IT) organizations taking greater responsibility for the management and support of *personal digital assistant* (PDA) deployments, IT staff have begun to assess the implications of PDAs within the overall context of their computing infrastructure. The PDA invasion of corporate America has found a much more prepared IT infrastructure than did the similar invasion of PCs during the late 1980s and early '90s.

When PCs entered the workplace as part of a decentralized proliferation, IT groups often believed their mandate was to support big iron or mainframe systems only. As a result, they often ignored the implications of the increasing number of PCs and their role in this new event until screen-scraping applications, or PC-based programs designed to simulate a user logged onto the mainframe at a terminal by hitting keys and then reading and storing the data presented, began to materially impact mainframe performance through what today would be referred to as a *denial of service* (DoS) attack.

Most IT groups today have a much broader and more holistic vision of their responsibility to safeguard access to corporate information and systems. As PDAs have begun to connect to IT-supported infrastructures, either through physical connections to desktop computers or network connections through wireless access points, IT staff have quickly assessed this as one of their core tasks and have begun to respond.

However, even though IT organizations have been able to respond to the PDA challenge, they must overcome the immaturity of the handheld computing movement in two primary ways. First, they must work with the handheld hardware and software community to address Enterprise and not just consumer requirements, which have historically been the main focus. Second, IT organizations must require professionals to read books like this and, in general, get up to speed on the unique and not-so-unique issues of managing PDAs.

PDAs: FROM CONSUMER TO ENTERPRISE SECURITY

As Enterprise IT organizations struggle to get better support from the handheld industry and try to stay on top of management and security issues, PDA security often slips by on the priority list. Although most organizations ensure that security has been considered prior to a major deployment, this consideration often becomes rushed and poorly planned. With the exception of those organizations that support dedicated security groups capable of evaluating PDA security products, the topic of security is usually overlooked until the late-stage of PDA deployment.

Whether a dedicated security group exists with the resources to anticipate and evaluate the issue of PDA security or not, we tend to have organizational problems. First, most groups evaluating PDA security often don't have a budget or adequate support to result in purchasing PDA security software and management systems for deployment to their existing PDA base. Second, if a security evaluation is being planned for a PDA deployment, most IT groups don't address PDA security until the late stages of the rollout. At this point, they generally have not properly budgeted for PDA security or allowed much time for the review and selection process.

The PDA security efforts that I am involved in take one of two forms. The first is a process, often led by a security group but sometimes led by an IT manager in charge of handheld policy, to set a standard for the organization. The other process is one that takes place after a large purchase of PDAs for mission-critical use, and at the last stage of the project someone asks, "What about security?"

Although this echoes the role security has historically played in computing, except for the government, military, and other limited areas, increasing the awareness of this topic will hopefully lead to an earlier integration of security into the PDA purchase process.

Evolution of the Security Challenge

Until somewhat recently, security was not a discipline that focused on the computing infrastructure. Network computing, probably more than any other historical shift, expanded security from the realm of physical security and into the digital age of the 1990s. However, the rapid proliferation of networked computing infrastructure and its continuingly rapid evolution have vastly outpaced the ability or willingness for corporate investment. As a result, the security discipline operates as an area always struggling to provide the training, policies, and infrastructures needed to confront the evolving risk-management challenges faced by modern organizations.

Physical Security in the Networked Computing Age
I always loved how the networked computing age took a group of people, often appropriately labeled as computer geeks, and turned them into computer jocks. When I worked in the area of *local area network* (LAN) technology in the early 1990s, I worked alongside characters that were the image in full. For the most part, however, computer guys didn't change; we just became cooler. When the Internet hit in full force, I even reached a point where my wife, usually embarrassed to discuss

my work at parties, began proudly announcing that I worked as a technical person in the Internet area.

Although I thought the Internet bubble was going to be the height of my professional cool with her, I found that the security thing, which became high profile while I was working to secure banking networks, has a new kind of cool. With the high-profile hacking stories in the press and the hint of cops and robbers, the computing security field offers IT professionals a new chance at cool.

The networked computing security discipline, however, requires many of the same elements that have been a part of traditional security professions. Although IT professionals may enjoy the "cops and robbers" excitement of incorporating the "Security" word into their title or job functions as a new designation, as we explore the topic of PDA security, we must be vigilant to consider a broad view of the security challenge and realize this is not something addressed with the simple purchase of a software package. Security requires understanding the nature of threats, designing policies and practices to remediate them, and then ultimately incorporating technology to enforce the will of an organization. As we define the policies and practices for mitigating risk.

During the course of this book, we will spend time assessing risk, but much of our focus will be on technology. As we explore the nature of brute-force password attacks or the technical nature of encryption, remember the risk of a help desk without strong policies that prevent the accidental release of a password for a lost device. As my friend and security mentor Rebecca Bace reminds me on many occasions, social engineering such as a simple phone call to an average user to get a password, not technical wizardry, remains at the heart of the most dangerous security threats.

From Networks to Mobility As organizations responded to the threat raised by public networks connections, they scrambled to put firewall services and later intrusion detection and prevention tools in place. As a profession, security largely focused on building a strong perimeter to protect infrastructure assets from the outside world. I have had the good fortune to work with world-class professionals in the area of network security management. In my experience, the best professionals never lost sight of the broader context or the security standards that require a strong policy to manage social engineering risks. In short, the Social Engineering threat reflects the fact that human beings with all their inherent limitations, play a key role in our organizations.

Although we established strong perimeters for our infrastructure, the highly mobile PDAs, first introduced with corporate approval as personal information management devices, were subject to risks that could bypass our strong perimeter and compromise our organization. It is the same risk faced by laptop computers: being easily lost or stolen. A study done by Safeware, The Insurance Agency, Inc., reports that 591,000 laptops were stolen in 2001. With respect to PDAs specifically, the Gartner Group estimates that 250,000 handheld devices were lost in U.S. airports in 2001.

In addition to PDAs' risk of being lost or stolen, the problem is compounded by the fact that people tend to use the basic *personal information manager* (PIM) applications to store sensitive information, such as network passwords. Suffice it to say, PDAs have introduced a new class of risk into the overall challenge of risk management within an Enterprise, and Section Two, *"Handhelds in the Enterprise: When, What, and How,"* explores a thorough discussion of this topic.

An Organization's Interest in Security Varies by Industry

All Enterprise exposure to lost or stolen PDAs is not created equal. Affected both by usage and the nature of the industry, some organizations clearly have a lower risk than others when it comes to PDA security. As we will discuss in Section Two, assessing the risk of an organization requires evaluating a number of factors.

In working with different organizations, I have healthcare companies and government entities express a much stronger interest in PDA security than anyone else. This interest in security not only results from the specifict types of business software or technology in use, but also as a result of industry-level issues. The best example of industries that show a stronger-than-average interest in PDA security are those that are directly regulated to meet security requirements.

Role of Regulation   Although public companies as a whole are evaluating how to respond to the financial disclosure/reporting requirements of the Sarbanes-Oxley Act in the wake of Enron, industry-specific regulations have also had a substantial impact on how organizations address the question of security. In the healthcare industry, patient privacy provisions around digital information in the *Health Insurance Portability and Accountability Act* (HIPAA) and financial information privacy issues in the financial services industries Gramm-Leach-Blily regulations have forced organizations to examine how they manage their information and the associated technical infrastructure like PDAs that contain it. Several industries, including healthcare, financial services, and government, have become regulated with respect to their public responsibility to protect information.

Although the initial intent of these statutes may focus on mainframe computers and data repositories that keep large amounts of confidential consumer information, this regulation on how electronic information is treated directly impacts PDAs. Networked distributed computing has made access to an organization's information available to the smallest and least managed technology components within an organization. However, regulation will not make organizations exempt from their responsibilities to safeguard that information because of the medium upon which it is stored.

Healthcare   In February 2003, the U.S. Department of Health and Human Services issued the final adoption rules for the HIPAA, which directly addresses healthcare organizations' responsibilities to safeguard electronic patient information they hold in trust. These regulations affect the entire industry and not just the organizations that directly deal with patients. Whether the company is a medical manufacturer storing information on which patient received an artificial heart, or an independent doctor getting a referral from an HMO, everyone entrusted with patient information is responsible.

Financial Services   Financial services, similar to healthcare, maintain a great deal of proprietary information regarding consumers. Rather than health-related, they hold financial information, which some might argue should be safeguarded more aggressively. Gramm-Leach-Blily reflects a more limited form of legislation, addressing how financial service companies or chartered banks deal with their customers. However, beyond legislation, financial service firms also differ from many industries in their strong cultural focus on the responsibility of safeguarding information.

Legal   The legal profession is another industry that has a strong history of addressing the topic of privacy of information. PDA management has simply become a natural extension to the questions of confidentiality that have guided their organizations largely since inception.

Government and Military   Government and military organizations have a long history of struggling to manage information. Balancing the right to public access and the managing of secrets, this area has reflected a clear focus on addressing the policy enforcement for PDAs.

Vendors Addressing PDA Security

Despite the immaturity of the hardware and software industries that are focused on the Enterprise-specific needs of PDAs, a segment of the market has begun to focus not only on PDAs, but increasingly on the PDA security area as well. Figure 2-1 illustrates the various hardware and software segments expected

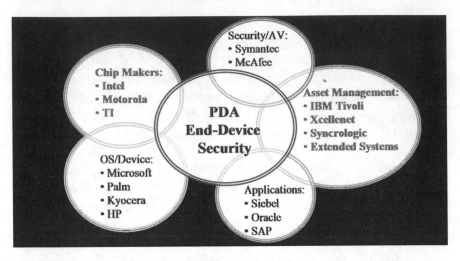

FIGURE 2-1.   Market segment affecting PDA security.

to increasingly adopt PDA security to meet the needs of Enterprise customers.

OS and Device Providers   *Operating system* (OS) and device providers have always provided some level of security in their device implementations. As will be further discussed in later chapters, both providers, the OS provider especially, have a strong interest and responsibility in ensuring adequate security. As a result, as the market for PDAs and security matures, OS and device providers are likely candidates to provide not only Enterprise-class, on-device security enforcement solutions, but also the libraries and software/hardware interfaces to enable extensible security solutions.

As the primary providers of PDA OSs, key players such as Microsoft, Palm, Symbian, and *Research In Motion* (RIM) illustrate the role we can expect to see the OS play in on-device security. RIM, in an effort to directly address the Enterprise security market, provides the most compelling example of how OS providers will directly engage in security, but their integrated multiple-device strategy still falls short of fully securing the device. Microsoft and Palm have illustrated their commitment to providing cryptography and other libraries for providers to extend their OS security solutions, but until recently they had not gone beyond a simple authentication based on *personal identification numbers* (PINs). However, moving forward, OS providers have increasingly provided strong leadership and are active advocates for strong security.

Device providers ranging from Palm to the recent success of Dell, with their Microsoft-based Pocket PC release, have also been challenged by the security issue. As a direct seller to Enterprise organizations, they daily face the question of security from Enterprise buyers. This challenge has led some end-device providers to seek support from existing security providers, such as Hewlett-Packard, to pioneer security solutions, such as the fingerprint reader illustrated in Chapter 3, *"The Power Resource Guide to Understanding Where Security Must Be Achieved."* End-device providers will increasingly play an integral role in the direction of PDA security. This is due to their often direct sales relationship with the Enterprise and their general role of identifying the final build of software to be incorporated into a specific device.

Much of the strong commitment that continues to emerge from hardware and software providers will continue to result from an overall desire to boost device sales in the Enterprise market. However, security, like any general asset management challenge, is a key barrier to the expansion of device sales to Enterprise organizations that would have to support thousands of devices within their existing infrastructure.

Chip Makers   Similar to OS and device providers, chip makers are part of a technological ecosystem that is primarily motivated by a drive to sell more devices. Since Texas Instruments, Motorola, and Intel represent the *central processing units* (CPUs) in almost all PDAs sold, if a lack of security slows the overall market growth, they will take an interest in increasing the availability of security solutions. As Enterprise purchasing increases as a relative percentage of overall device sales, the needs of Enterprise organizations will increasingly drive not only the actions of the end-sellers of devices, but the whole PDA market, including chip makers.

Security and Antivirus Providers Security providers like Symantec and Network Associates have worked to pioneer software clients to run on the desktop as agents constantly operating in the background. These providers have expanded their mandates to address broader threats to the computing device, including but not limited to firewall and intrusion-detection services. This expanding strategy to protect the end-device from external threats integrates PDA end-device security with an overall security solution that includes authentication, encryption, and other various components.

PDA Killer Applications A series of killer business applications have emerged to drive sales of PDAs within the Enterprise. If these solutions are to reach critical mass, issues including software distribution, asset management, and security are going to require solutions. Importantly, these killer or mission-critical applications tend to also bring proprietary information or functionality from the safety of the corporate network to the relatively high-risk domain of the mobile PDA. These factors and others have lead to the sale of security software designed to turn the PDA from a consumer device into a mission-critical tool for an organization.

Siebel's mobile *customer relationship management* (CRM) software is a key example of a solution driving many of the corporate PDA rollouts I am currently working on with Enterprise clients. However, once a healthcare organization, for example, deploys a mobile application that might have patient information on it, they have a legal responsibility to provide the necessary security to safeguard that information.

Asset Management Probably the best example of an area that has directly responded to an Enterprise's needs with respect to handheld computing is the area of asset management. In IT's management area for servers, desktops, and laptops, HP OpenView and IBM Tivoli continue to define key Enterprise software to support the distribution of software as well as various audit and control capabilities for the management challenge.

Although the PDA rollout faces many of the same challenges, they have recently received the dedicated support of organizations dedicated to the challenge of asset management. A number of companies, including Xcellenet, Syncrologic, Extended Systems, and IBM Tivoli, provide agents to operate on the PDA device, enabling control of software distribution and audits, which have come to be an integral part of the desktop world. These companies, as illustrated in Chapter 3, support the emerging requirements of handheld computing. As the requirements continue to evolve, the asset management area will be a key driver of security requirements and how to manage, audit, and enforce them.

PDA SECURITY: LET'S GET STARTED

As a final, general orientation to the PDA field, Chapter 3 provides an overview of the OS providers, some sample software categories, and a comprehensive look at the device manufacturers in the emerging PDA market. Following Chapter 3's reference guide, we continue tackling PDA security from an assessment, policy, and enforcement perspective. We begin with a business-level look at the topic and move into a technical discussion of the key topics from encryption and authentication to hacking the PDA device.

THE POWER RESOURCE GUIDE TO UNDERSTANDING WHERE SECURITY MUST BE ACHIEVED

In addressing handheld security, just as in the broader discipline of security, understanding the environment to which your solution will be applied is critical. For handheld security, this means the hardware implementation, *operating system* (OS), and other installed applications operating with administrative privileges on the device. In later chapters, we will take a more technical look at the major environmental components; however, at this point, we provide a high-level introduction to the relevant software and hardware players in the *personal digital assistant* (PDA) field. This chapter outlines examples within each of the major PDA software and hardware category areas that must be considered as part of a security solution.

OPERATING SYSTEMS (OS's)

First and foremost, a security implementation relies on and contends with the capabilities of an OS. The OS of a device, more than any other single component, sets the rules and boundaries for executing security on the device. The OS defines the scope of functionality, both providing limits and exposing opportunities to would-be attackers and security vendors.

Within the PDA field, several players must be considered when it comes to the range of devices and software available for today's PDA. When considering the OSs on which these devices run, five key vendors emerge in the rapidly maturing market. Their OS offerings provide power for devices from the converging Smartphone field, such as Symbian, or manufacturer-driven OSs, such as Sharp's Linux implementation. However, the three primary operating environments to address from a market share perspective (92 percent of the combined worldwide share in 2002) include Palm OS, Pocket PC, and devices based on *Research in Motion* (RIM).

Palm, Inc. and PalmSource

We will begin by looking at a brief history of Palm. Palm, Inc. was initially formed in 1992 when the personal computer was just beginning to become a possibility for Mr. and Mrs. John Q. Smith. In 1995, U.S. Robotics acquired Palm, Inc. and released their first devices in 1996. These devices can be considered the first PDAs. However, the Palm Pilot 1000 and 5000 were not the first devices to be called PDAs, but rather were what many consider to be the birth of the modern PDA.

One year later, in 1997, U.S. Robotics was acquired by 3Com, thus making Palm, Inc. a subsidiary of 3Com. Palm continued to refine and work on both the Palm OS as well as their hardware, offering better displays and more refined personal information management software. In 2000, 3Com decided to launch Palm, Inc. as a stand-alone company. Following this separation and after two successful years of developing software and hardware for the Palm OS, Palm decided to separate its software division from its hardware group.

In 2002, Palm, Inc. became the parent of two increasingly independent companies: Palm Source and Palm Solutions Group. Palm Source handled the development, licensing, and distribution of the Palm OS and other software, while Palm Solutions Group continued to create and sell hardware solutions utilizing software created by Palm Source. At this time, the Palm OS was placed on devices from several other manufactures, most notably Sony, which has taken an equity interest in Palm Source and Handspring. We'll further discuss these manufacturers later in this chapter.

Palm remains the market leader when it comes to both devices and OSs. As of April 2002, the Palm OS is being used by 57 percent of PDAs, giving it a clear lead (IDC, *Interactive Data Corporation*, Report, April 2002). According to Palm, their current installed base of devices is approximately 24 million (according to Palm Source information presented at October 2002 investors briefing).

Devices powered by the Palm OS range from PDAs to phones to even watches. With its early market leadership, the Palm OS has proliferated a wide range of devices that have found their way into the hands of everyone from college students to *information technology* (IT) professionals. Doctors use them when gathering patient information, while lawyers store legal briefs on them, rather than carry a stack of papers.

Microsoft's Pocket PC 2000, 2002, and .Net

The next stop on our tour of the major players in PDA OSs is the software giant Microsoft. We all know Microsoft as a maker of desktop OSs such as Windows. Their first venture into PDA OSs was Windows CE, which was introduced in 1996. It was made to look very much like Windows 95 in the way applications and the OS interface behaved. Microsoft hoped that users would need little time to adapt to their Windows CE OS from the Windows 95 platform, which had such a mass market adoption in the desktop world.

Windows CE 1.0 was essentially a slimmed-down version of Windows 95 designed to fit on a handheld device. At the start, Windows CE did not handle its versioning as its desktop brethren, but eventually made it through 2.0 up to 3.0. Windows CE 3.0 is what is used today.

Sitting on top of Windows CE 3.0, extending the available *application programming interfaces* (APIs) and other features, is Pocket PC. Initially, we began with Windows CE 3.0 Pocket PC 2000, and new devices today are still running it. Microsoft is currently in the process of upgrading Windows CE, which will move Pocket PC forward onto a .Net version that will be widely available. Unlike Palm, Microsoft does not make its own devices, which are powered by the Pocket PC OS, but rather they rely on third-party vendors such as Compaq and Dell. We'll discuss these vendors more later.

Devices powered by the Pocket PC OS hold the second largest market share behind Palm. According to the April 2002 IDC report, Pocket PC devices hold a 21 percent market share. These devices tend to be more expensive than their Palm rivals, but this is mainly because the Windows CE Pocket PC OS has historically targeted users that require more RAM, faster *central processing units* (CPUs), and better screens. With the recent release of the Dell AXIM devices at the Pocket PC low end, combined with Palm Tungsten devices at the Palm high end, these pricing distinctions are decreasing.

Linux and Java

Currently, only one real manufacturer, Sharp, is designing devices that are meant to run Linux. We'll talk more about the devices Sharp is currently making later. Linux has found its way onto almost every PDA on the market. Advanced users have taken the packaged OSs off their devices and replaced them with Linux. From Palm to Handspring to iPAQs, you name it, Linux is there. You can also find devices based on Linux in many specialized roles, such as handheld barcode scanners.

Linux provides a robust, open-source alternative to the Palm or Windows CE OS. The downside to these devices is that only a small field of applications is available from which to choose, unlike Palm and Pocket PC-based devices. With that said, Linux devices work hard to provide support for a *Java Virtual Machine* (JVM) that enables them to run Java-based applications, which are becoming increasingly available and are easily developed.

BlackBerry RIM OS

Devices based on the RIM OS rely on an always-on network connection to provide constant access to the Internet and email. They also offer the standard personal information management software that all PDAs have. These items include applications such as a calendar with appointments, contacts, and to-do items. Rather than going with the conventional stylus, RIM devices offer a built-in thumb keyboard in the standard PC layout. A jog dial, which is a rotary/button navigation tool, is provided for navigating the system. This allows for fast and easy typing as well as instant messages over a *standard management system* (SMS). The reliance on the thumb keyboard has set a standard that other emerging email-centric devices emulate.

The latest incarnations of RIM devices utilize the latest Java 2 Micros Edition, which gives them extensive versatility for which previous OSs have not been known. Because of their always-on connection, the new RIM devices make an excellent platform for applications that rely on an Internet connection, such as instant messengers. The latest version of the OS also enables devices to incorporate a built-in phone capability, allowing this OS to offer the ultimate in connectivity.

Symbian OS

We hear a lot of talk about Smartphones and convergence, or essentially the merging of PDA capabilities with cellular phone functionality. As described previously, convergence devices tend to be primarily PDAs with phone capabilities versus devices that are primarily phones with PDA capabilities. In the latter category, these have become synonymous with the term Smartphone, and although the battle for dominance rages on, Nokia with its focus on Symbian leads the way.

The Symbian OS was designed from the ground up to work on Smartphones. Symbian, an outgrowth of Psion, was brought to the phone market when it was spun into a joint venture in 1998 by Ericsson, Motorola, Nokia, and Psion. The Symbian OS was designed to have a small footprint, or resource requirement, on the SmartPhone device. The first Symbian-based device was shipped in 2000, but several others are now on the market and many more are being prepared from such companies as Nokia and Sony Ericsson.

Because of the Symbian OS's versatility on Smartphone devices, device designers who want to use the Symbian OS have little to no limitations. When making Smartphone devices for Pocket PC or Palm devices, certain restrictions must be met. For example, Palm and Pocket PC devices must have a certain pixel-by-pixel size for their screens, whereas Symbian OS devices have no such restrictions. This can all translate into a faster time to market as well as an OS evolution. Symbian is the dominant platform for Smartphones, and in Europe this is primarily due to the reach of Nokia. However, most of the devices have yet to support the types of Enterprise applications that seek security.

CORE APPLICATIONS

Application software layered onto the OS provides both the tools for achieving security and an introduction to the challenges and threats that must be managed. The following section explores key categories within our focus OSs of Microsoft and Palm, and it addresses the agent-based system management software that enables the asset management of the handheld devices within an organization.

The brief overview covers built-in software, IT administration, *virtual private networks* (VPNs), on-device security, and tools that store and manage data. These categories will demonstrate at least one example, but it should be understood that they are just a sample representation of a broader set of solutions. Additional categories include general hacks as well the broad range of utility and productivity solutions that can be found for every OS.

Built-in Software

All devices will be shipped with a factory default image, or software load-out, burned on the *read-only memory* (ROM). This image can be restored at any time with a hard reset of the device and represents the set of base applications with which the device was shipped. These applications generally start with the OS provider and are then extended by the final distributor or manufacturer of the device. The following provides an example of the software implemented by the OS provider; however, each manufacturer or end distributor can and often does extend this offer to differentiate themselves in the marketplace or to implement some unique aspect of their device.

Palm Built-in Applications   Every Palm OS-based device comes with a set of built-in applications, and most of them are related to *personal information manager* (PIM) applications. However, hardware manufacturers, especially licensers of Palm Source, can substantially extend the default set of applications. At a least common denominator level, the following applications are provided:

- Address book
- Date book
- To-do list
- Memo pad
- Mail
- Hotsync
- Note pad (Palm OS 4.0 only)
- Expenses (Palm OS earlier than 4.0)
- SMS (Palm OS 4.0 from add-on CD)
- WebPro (Tungsten-T only, from add-on CD)
- VersaMail (Tungsten-T only, from add-on CD)

These applications provide for almost all the personal information management requirements, but well over 10,000 third-party applications have been developed to extend Palm

OS-based devices. A directory from the Palm Solutions Group of additional applications can be found at www.palm.com/software.

Microsoft Built-in Applications   Although not as extensive as the more than 10,000 applications available for the Palm OS, Microsoft has seen third parties develop a wide variety of solutions for their WinCE/Pocket PC OS, due to Microsoft's commitment to development tools and increasing market penetration. The following section targets several categories and examples, and a more complete list of available solutions can be found at www.microsoft.com/mobile/enterprise/solution-providers.

The Pocket PC/Pocket PC 2002 device is shipped with a set of pre-installed software. The set of applications can vary somewhat from manufacturer to manufacturer, as they add additional *original equipment manufacturer* (OEM) offerings to differentiate their product, but at a minimum the following is included:

- ActiveSync
- *PIM* (Pocket Outlook)
  - Contacts
  - Calendar
  - Tasks
  - Mail
  - Notes
- Office
  - Microsoft Pocket Word
  - Microsoft Pocket Excel
  - Microsoft Reader
  - MSN Messenger
  - Terminal Services Client

- Multimedia
  - Microsoft Media Player
- Games
  - Solitaire

IT Administration

As a category, this group can include hack programs and can represent a threat to the proper operation of a custom application. The two examples here illustrate the general-feature areas or capabilities you can readily find to support your efforts to troubleshoot and manage devices. These tools can also be useful in evaluating the quality and strength of security and management tools of various kinds.

File Z

Manufacturer: Nosleep software (freeware)
Product name: File Z
Requirements: Any Palm OS hardware
The File Z features include:

- Used memory/total memory display.
- Menu option to set the backup/copy bits of all files at once.
- Edit the create/modify/backup date/times (including the Never option).
- Customizable second column: creator, type, size, record count, attributes, creation date, modified date, and backup date.
- Many filtering options: filename, creator, type, size, record count, and attributes.
- Shows files on external memory cards.
- All file attributes can be changed.
- Hide-ROM option.

- Supports MemPlug cards.
- Multiple file delete/move/copy capabilities.
- Preference editor.

MobiServer

Manufacturer: Northern Parklife, Inc., www.northern.net
Product name: MobiServer, www.northern.net/mobiserver
Requirements: MobiServer enables the remote administration of a computer in the Enterprise network via any device enabled with the *Wireless Application Protocol* (WAP), including Pocket PC 2002 and Pocket PC Phone Edition-based devices.

The MobiServer features list is as follows:

- Ping for testing connectivity to selected computers.
- Get system info.
- Event log.
- Task manager.
- Service control.
- File manager.
- User manager.
- Popup user messages.
- Reboot.
- Quota management.

SonicAdmin

Manufacturer: Sonic Mobility, Inc., www.sonicmobility.com
Product name: SonicAdmin
Requirements: Available for Pocket PC and RIM devices
The SonicAdmin features are as follows:

- User management.
- Event logs.

- *Domain name system (DNS)* management.
- File manager.
- Email capabilities from controlled servers.
- Services monitor.
- Command line, Telnet, and *Secure Shell (SSH)*.
- Server management with reboot, power control, and more.
- Statistics.
- Workthrough secure connections.
- Support of Windows and UNIX servers.

Virtual Private Network Support

An often misunderstood area, *virtual private network* (VPN) solutions generally do not provide any on-device security and rather focus only on securing information in transit to or from another device. With the expansion of wireless-enabled devices, VPN solutions have come to play an increasingly important role in an overall security policy.

MovianVPN
   Manufacturer: Certicom Corp., www.certicom.com
   Product name: MovianVPN, www.certicom.com/products/movian/movianvpn.html
   The MovianVPN key performs the following tasks:

- Integrates with leading VPN gateways.
- Uses *IP Security (IPSec)* and incorporates *Elliptical Curve Cryptography (ECC)*.
- Establishes a secure encrypted tunnel between Pocket PC devices and Enterprise networks.
- Supports two-way authentication systems such as SecurID.
- Provides GSE versions with *Federal Information Processing Standards* (FIPS) 140-2 certification.

VPN-1 Secure Client

Manufacturer: Check Point Software Technologies Ltd., www. checkpoint.com

Product name: VPN-1 SecureClient, www.checkpoint.com/products/connect/vpn-1_clients.html

The VPN-1 Secure Client performs the following tasks:

- Integrates with VPN-1 Pro Gateway.

- Supports IPSec.

- Supports *Data Encryption Standard (DES) and Triple DES (3DES)* algorithms.

- Supports *public key infrastructure (PKI)*, such as RSA and Diffie-Hellman algorithms up to 1,536-bit keys.

- *Microsoft Common Application Programming Interface (CAPI)* support.

- *Lightweight Directory Access Protocol (LDAP), New Technology (NT)* domain, and active directory support.

Mergic VPN

Manufacturer: Mergic, Inc., www.mergic.com

Product name: Mergic VPN for Palm OS v1.03

Mergic VPN establishes VPN connections from Palm handheld devices to Enterprise networks. Its key features include the following:

- Point-to-Point Tunneling Protocol (PPTP) support.

- Auto-connect features.

- VPN connection indicator.

- Support for 40- to 128-bit *Microsoft Point-to-Point Encryption (MPPE)* keys.

- Support for the *Password Authentication Protocol (PAP)*, the *Challenge Handshake Authentication Protocol (CHAP)*, and the *Microsoft Challenge-Handshake Authentication Protocol (MS-CHAP)* versions 1 and 2 authentication.

- High-optimized encryption.

Databases

In achieving on-device security, the information to protect must be well understood in both structure and location. In addition to document metaphors in both Palm and Pocket PC, popular database applications leverage the file systems to store complex binary structures in order to manage data. In addressing encryption and other file-level security controls, these applications should be well understood and be potentially directly addressed in a solution.

Microsoft SQL Server CE

Manufacturer: Microsoft Corporation, www.microsoft.com

Product name: Microsoft SQL Server CE 2.0, www.microsoft.com/sql/CE/default.asp

Microsoft's *Structured Query Language* (SQL) Server 2000 Windows CE Edition (SQL Server CE) version 2.0 is a compact database for rapidly developing applications that extend Enterprise data management capabilities to mobile devices. SQL Server CE makes it easy to develop mobile applications by supporting SQL syntax and providing a development model and API consistent with SQL server.

SQL Server CE has the following advantages:

- *A familiar database platform for rapid development.*

- *A compact yet capable relational database.* SQL Server CE has a small memory footprint, delivering all its functionality in approximately 1 MB. Performance is enhanced with an optimizing query processor. A range of data types is supported to ensure flexibility, and 128-bit encryption is provided on the device for database file security.

- *Flexible data access.* Remote access is provided for data in SQL Server 6.5, SQL Server 7.0, and SQL Server 2000 databases through the remote execution of Transact-SQL statements and the capability to pull record sets to the client device for updating.

DB2 Everyplace

Manufacturer: IBM Corporation, www.ibm.com

Product name: DB2 Everyplace Database, www-3.ibm.com/sfotware/data/db2/everyplace

DB2 Everyplace is available in two editions—Database and Enterprise. It supports several platforms:

- Palm OS
- Symbian OS version 6
- Microsoft Windows CE/Pocket PC
- Win32 (Windows NT and Windows 2000)
- QNX Neutrino, Linux, and embedded Linux devices

HanDBase

Manufacturer: DDH Software, Inc., www.ddhsoftware. com/

Product name: HanDBase, www.ddhsoftware.com/handbase_palm.html

The HanDBase features include:

- Visual form creation.
- Support of various data types.
- Peer-to-peer synchronization.
- Quick search.
- Viewer set support.
- Encryption support.
- Access permissions.
- Calculation fields support.
- Links between databases.

Document Management

In contrast to relational data structures illustrated in the previous database packages, document-centric representation, using both built-in file systems for document storage and binary structures to index and manage documents, are exposed to the user as simple file systems. Again, these applications must be understood to determine whether they must be directly addressed in an overall security strategy.

File Explorer The file explorer capability is not illustrated under the built-in applications because of its tight integration with the OS. It is only mentioned here to illustrate the category of information that can be stored on the device outside the direct influence or control any specific associated application. Pocket PC provides the file explorer method most people are familiar with from using desktop computers, as it enables users to manage a wide array of documents.

Documents To Go

Manufacturer: DataViz Inc., www.dataviz.com

Product name: Documents To Go v5.0, www.dataviz.com/products/documentstogo/index.html

The Documents To Go package consists of several applications:

- *Word To Go*   Works with Microsoft Word documents.
- *Sheet To Go*   Works with Microsoft Excel documents, including charts.
- *Slideshow To Go*   Enables you to work with Microsoft PowerPoint documents.
- *DataVizMail*   Provides email capabilities.
- *PDF To Go*   Enables you to open and view PDF files.
- *Pics To Go*   Views graphic images.

Quick Office

Manufacturer: Cutting Edge Software, Inc., www.cesinc.com

Product name: Quick Office Pro 6.2, www.cesinc.com/quickoffice_pro

QuickOffice Pro includes several components:

- *QuickPoint*   Views/edits Power Point presentations.
- *QuickWord*   Views/edits Word documents.
- *QuickSheet*   Views/edits Excel spreadsheets.
- *QuickChart*   Works with Excel charts.

WordSmith

Manufacturer: Blue Nomad LLC, www.blue-nomad.com
Product name: WordSmith 2.2
WordSmith is full-featured Word document editor with the following features:

- A full-featured word processor, free electronic book viewer, and enhanced memo pad.
- Capability to read and edit documents directly from a *Secure Digital* (SD)/*Multimedia Card* (MMC) or Memory Stick.
- Seamless integration with Microsoft Word (Windows only).
- Spellchecker and Thesaurus included.
- Bidirectional conduit for Windows and Mac OS, and command-line converters for Windows and Linux/Intel.
- Capability to print documents with PrintBoy from Bachmann Software.
- Standard functions such as Cut, Copy, Paste, Multipaste, Select All, Undo, Redo, and Find.
- Font-style support.
- Beam, Delete, Duplicate, Rename, Save, and Save As functionality.
- Paragraph formatting, such as setting page breaks, indents, and line spacing.
- Support for bulleted lists.
- Fast compression of documents during HotSyncs and on demand.

- Auto-scrolling, pen scrolling, and special teleprompter-style scrolling for easy viewing of documents.
- Capability to synchronize documents to multiple computers.

Agent-Based Distribution/Management Software

Handheld device management and distribution solutions form the natural extension to the mature asset-management software segment that developed to address the same problem with desktop and later laptop computers. IT administrators have always struggled with managing software distribution, maintenance, administration, and audit control issues associated with distributed computing devices. Although HP OpenView and IBM Tivoli's names are synonymous with asset management in the desktop space, a new set of companies has emerged to play a role with handheld devices. The following four companies, including IBM Tivoli, illustrate the driving solutions to support handheld system management in the marketplace today.

XTNDConnect Server
    Manufacturer: Extended Systems
    Overview: Extended Systems XTNDConnect Server supports data distribution and a wide array of management support. Device support is provided for Palm and Pocket PC, among others.

Afaria
    Manufacturer: Xcellenet
    Overview: Afaria addresses mobile device management as well as the distribution of data. Device support includes Windows, Java, Symbian, Windows CE/Pocket PC, Palm, and BlackBerry.

IBM Tivoli Configuration Manager
    Manufacturer: IBM Tivoli
    Overview: A distribution and management solution, it supports Palm OS, Pocket PC, and Nokia Communicator devices.

RealSync Server
Manufacturer: Syncrologic
Overview: Like the others before Syncrologic, it too supports data distribution and a wide array of devices.

Security

The broad category of security here represents on-device protection of the PDA. This category is the focus of the next section of the chapter and PDA defense is the basis of the case study in Chapter 6, *"The How-to Guide: Asynchrony's PDA Defense, Palm's Choice for the Enterprise."* The on-device security solution generally becomes key when achieving strategic policy management goals for the handheld devices within an Enterprise.

PDA Defense
Manufacturer: Asynchrony Solutions, www.asolutions.com
Product name: PDA Defense, www.pdadefense.com
Requirements: Palm OS, Pocket PC, RIM devices
Key PDA Defense features are as follows:

- Strong alphanumeric password use.
- Automatic device locking.
- Automatic transparent encryption of PIM data and files.
- Transparent encryption of the data stored on the external memory cards.
- Support of virtual encrypted volumes.
- Application launch protection.
- Hardware button password entry.
- Selective data bit wiping.
- Strong Blowfish 128-bit and 512-bit key encryption.
- Centralized administrative security policy management.
- Compatibility with XTNDConnect Server, ScoutSync Server, AvantGo Enterprise, and Afaria by XcelleNet.

MANUFACTURERS

Manufacturers are an important element in the success of PDAs. No one would want to use a PDA if it is clumsy, bulky, and difficult to transport, no matter how well the OS provides application functionality. Most of the OSs discussed have several different manufacturers that produce devices based on these OSs. We will begin by looking at the different manufacturers of devices that utilize the Palm OS.

Palm OS-Based Devices

Palm Solutions Group, as a subsidiary of Palm, Inc., has been making devices for the Palm OS longer than any other manufacturer. Their first device was released in 1996 with the Palm Pilot 1000 and 5000. Since then, they have created better devices incorporating advances such as wireless communication, color backlit *liquid crystal displays* (LCDs), and external memory card readers (see Figure 3-1).

Palm Solutions Group manages to hold the largest market share of Palm OS-based devices within the world-wide market. Because the group is part of the original company that created the Palm OS, the majority of consumers see Palm Solutions Group as the primary source of Palm OS-based devices. They

FIGURE 3-1.   Palm OS-based device.

offer a wide range of devices from the entry-level Zire, now priced below $99, to the Enterprise-oriented Tungsten line of devices, including a range of wireless capabilities.

Another major manufacturer of Palm OS-based devices is Sony. As the first company to take an equity position in Palm Source, they have a strong commitment to the Palm OS and have recently introduced a successful form factor for the Palm OS in their Clie NV and NX series devices. These devices, with their built-in keyboard that can be swiveled out for easy access, have established a novel design in which consumers have taken a strong interest.

The next major player when it comes to Palm OS-based devices is Handspring. This company was founded by some of the original founders of Palm. They left Palm in order to create a new company whose primary concern was to make devices based on the Palm OS. One line of devices that has done well for Handspring is their Treo. Currently, four such devices are available. Three of them are in the Smartphone category, making them a hybrid of phone and Palm PDA, while the fourth is a standard PDA that looks just like its Smartphone counterparts. The Treo devices depart from the norm of Palm OS-based devices and replace the handwriting recognition area—with a hardware keyboard, much in the same way as the BlackBerry RIM devices. However, unlike the RIM devices, the Treo devices still rely on a stylus.

Another Smartphone device creator for the Palm OS is Kyocera. They released a popular phone, the Kyocera 6035. This was basically a Palm device with a flip cover that provides a dialing pad. Once opened, it reveals a standard-looking Palm. Many consider this to be the first successful Smartphone device. The new 7100 series Smartphone product from Kyocera incorporates a clam shell design and is much smaller than the original 6035 model, including a *Secure Digital* (SD) card slot, MP3 player capabilities, a color screen, and more.

Several other companies offer devices that use the Palm OS. These devices include standard PDAs from a long list of companies, providing form factors ranging from the standard Palm style to watches from Fossil, Inc. The following is a sample list of these companies:

- The Acer Group
- Alpha Smart, Inc.
- Garmin
- GSL
- Legend
- Samsung Electronics
- Symbol Technologies

The remainder of the Palm manufacturer discussion focuses on providing a more detailed look at the specific devices and functionality provided by key manufacturers.

Palm Solutions Group Device Characteristics   In addition to the devices listed in Table 3-1, Palm plans to release new phone-enabled models in 2003, including the Tungsten W, which has a built-in mobile phone that uses the *Global System for Mobile Communications* (GSM) and the *General Packet Radio Service* (GPRS) with Palm OS.

Palm by Any Other Name: Handspring, Sony, and So On
Palm OS-based devices come in a wide variety of forms and configurations. They arguably provide strength in the ongoing fight with Microsoft for dominance. The following array of devices provides only a sampling of the Palm OS-based devices available in the market and further complicates the challenge of delivering a manageable security policy. Fortunately, the Palm OS forms the standard that security solutions can be developed within across a wide array of device types. Importantly, if your organization plans to purchase large blocks of devices, make sure you select and test the compliance of both your security and system management products prior to issuing the purchase order.

Table 3-1. Palm Solutions Device Summary

Feature	Tungsten T	i705	m515	m500	m130	m125	Zire	m105
CPU TI Operations, Maintenance and Administration Part (OMAP) (ARM)	✓							
CPU Motorola Dragonball		✓	✓	✓	✓	✓	✓	✓
Integrated wireless	Bluetooth	Palm.net						
Color screen	✓	✓	✓	✓	✓		✓	✓
Expansion card slot	✓	✓	✓	✓	✓	✓		
Rechargeable batteries	✓	✓	✓	✓	✓			
Palm OS	5.0	4.1	4.1	4.0	4.1	4.0	4.1	3.5
Infrared Data Association (IrDA)	✓	✓	✓	✓	✓	✓	✓	✓
Email and web browsing	✓	✓	✓	✓	✓	✓		✓
RAM	16MB	8MB	16MB	8MB	8MB	8MB	2MB	8MB
Flash ROM	✓	✓	✓	✓				
Color	✓		✓		✓			
High-resolution display	✓							

Acer   Acer has licensed Palm OSs and manufactures Palm OS-based devices, as illustrated in Figure 3-2, with English and Chinese language support. Actually, Acer sells an s10 model with English language, Chinese Simplified, and Chinese Traditional support. It has voice recording capabilities and PIM applications set.

AlphaSmart   AlphaSmart Dane is an interesting PDA, as illustrated in Figure 3-3. It uses a wide LCD and a keyboard, and it may be treated as a notebook computer even though it operates on Palm OS 4.1.

Fossil   Fossil has delivered the first watch PDA based on Palm OS or any standard PDA OS to the best of my knowledge, as illustrated in Figure 3-4. Whether it will be a hit with consumers remains to be seen. The Wrist PDA does not incorporate Dick Tracy's videophone functionality, but it does take a step in that direction. If nothing else, I had to include this for the gee-whiz factor.

Handera   Handera 330 and the discontinued TRG Professional Handera were the first to introduce a Palm OS-based device with a memory card slot, as illustrated in Figure 3-5. TRG Pro supported one Compact Flash slot, and now the Handera 330 supports two slots: Compact Flash and Secure Digi-

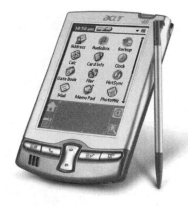

FIGURE 3-2.   Acer device.   **FIGURE 3-3.**   AlphaSmart device.

FIGURE 3-4. Wristwatch device.

tal. Also, HandEra LCD supports a high-resolution mode and a virtual graffiti area.

Handspring   Handspring produced a revolutionary product line: a Smartphone based on the Palm OS in the Treo series, as illustrated in Figure 3-6. As previously discussed, Handspring introduced four models:

- *Treo 300*   PCS network support and a color LCD display.
- *Treo 270*   A GSM/GPRS mobile phone and a color LCD display.

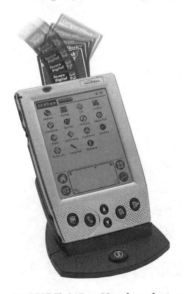

FIGURE 3-5. Handera device.   **FIGURE 3-6.** Handspring device.

- *Treo 180*   A GSM/GPRS mobile phone and a grayscale green LCD display.

- *Treo 90*   Doesn't have phone functionality, but it uses a Treo form factor and a color LCD display.

Kyocera   Kyocera manufactured a new model of Smartphone in the Kyocera 7135, as illustrated in Figure 3-7. Following its hit Smartphone, the QCP 6035 with Palm OS 3.5, Kyocera has raised the bar in convergence devices with the introduction of its first 7100 series. By including an SD card slot, an MP3 player, and a color screen on Palm OS 4.1, they have set the standard. Details about the product can be found at www.kyocera-wireless.com/7100_phone/7100_phone_series.htm.

Samsung   Samsung, following the lead of Kyocera, has worked hard to deliver a high-quality Palm OS-based *Code Division Multiple Access* (CDMA) Smartphone. Although the SPII-I300 picture in Figure 3-8 had reasonable success, the highly anticipated release of the I500 may leapfrog Samsung into a Smartphone leadership position. The I500 appears to closely mimic the features and functionality of the Kyocera 7100 series without the SD card slot. Separately, Samsung also plans to

FIGURE 3-7. Kyocera device.   **FIGURE 3-8.** Samsung device.

introduce revolutionary screen design and functionality with an upcoming WinCE device release that has great possibilities.

Sony   Sony, one of the largest Palm OS licensees typified by the example in Figure 3-9, manufactures a large set of different models illustrated in summary fashion in Table 3-2.

Sony offers the Clié handheld series with organizer and additional capabilities. Many of the Clié series handhelds include exclusive Sony features such as:

- Integrated keyboard.
- Built-in MP3 player.
- Jog dial (all models).
- High-resolution screen, 3203320 or 3203480, with 16-bit color support.
- Universal remote feature that enables the Clié handheld to be used as a universal remote control for a video recorder and TV set.

Symbol   Symbol Technologies produces a wide range of custom devices for industrial and government use that are often ruggedized, or otherwise adapted for nonstandard use. Some have built-in barcode readers or other hardware adaptations, and they are known for being big and bulky. Some of the devices

FIGURE 3-9.   Sony device.

Table 3-2. Examples of Sony Devices

	NZ90	NX70	NX60	T665C	SJ30	SJ20	SL10
CPU Intel Xscale PX250	✓	✓	✓				
CPU Motorola Dragonball				✓	✓	✓	✓
Bluetooth	✓						
Add-on wireless	✓	✓	✓				
Color screen	✓	✓	✓	✓	✓	✓	✓
Screen resolution	320×480	320×480	320×480	320×320	320×320	320×320	320×320
Expansion card slot Memory Stick	✓	✓	✓	✓	✓	✓	✓
Rechargeable batteries	✓	✓	✓	✓	✓	✓	✓
Palm OS	5.0	5.0	5.0	4.1	4.1	4.1	4.1
IrDA	✓	✓	✓	✓	✓	✓	✓
Digital camera	✓		✓	✓			
RAM	16MB	16MB	16MB	16MB	16MB	16MB	8MB
Flash ROM	✓	✓	✓		✓		
Color	✓	✓	✓	✓		✓	
Integrated MP3 player	✓	✓	✓	✓			
Jog dial	✓	✓	✓	✓	✓		
Integrated keyboard	✓	✓	✓				
Universal remote control	✓		✓	✓	✓	✓	✓

based on the Palm OS (see Figures 3-10, 3-11, and 3-12) include the following devices:

- SPT 1550.
- SPT 1700 (rugged device), a built-in 802.11b-compliant wireless adapter.
- SPT 1733/1734 (rugged device), a *cellular digital packet data* (CDPD)/GSM900/1800 MHz wireless modem.
- SPT 1800 (rugged device), a 802.11b-compliant wireless adapter and CDPD 800 MHz wireless modem.

Pocket PC-Based Devices

After Palm, Inc. broke through with the first widely adopted PDA, Microsoft entered the field with deep pockets and a strong commitment to the Enterprise marketplace. However, it wasn't until Compaq, now Hewlett-Packard, entered the market with the iPAQ line that Microsoft found a manufacturer with a form factor that could garner any kind of market penetration. Since Compaq's breakthrough device, the Pocket PC market has taken off.

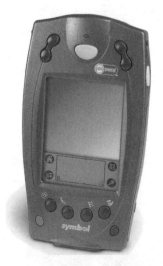

FIGURE 3-10. SPT 1700.   **FIGURE 3-11.** SPT 1550.

FIGURE 3-12.   SPT 1800.

One of the newest players in this arena is Dell. They have recently brought their Axim handheld device to market. The benefits to Dell's lineup of devices are their cost-per-features ratio. Dell offers a device with a fast processor and a lot of RAM while keeping the price as low as many Palm OS-based devices.

Additionally, Toshiba has recently made a foray into the high-end market with their latest offering. Their device, which is roughly the same size as an iPAQ, boasts both an SD memory card slot, a *Compact Flash* (CF) memory card slot, and built-in wireless 802.11b connectivity, whereas the iPAQ only offers an SD card slot standard. The top-end iPAQs also offer built-in Blue Tooth connectivity as well as 802.11b, but at a far higher price.

Several manufactures have also introduced Pocket PC Phone Edition devices. Two such manufacturers are Audiovox and T-Mobile. Both their offerings have many of the latest features one would expect in a high-end Pocket PC, as well as the capability to connect to the Internet to directly access email and other online applications. Samsung is currently working on a next-generation Windows CE .NET device that will employ Microsoft's increasingly Internet-centric approach.

Other manufacturers that create PDA devices based on the Microsoft Windows CE 3.0 Pocket PC OS include the following:

- Asus
- View Sonic
- NEC
- Casio
- Siemens
- Mexmal
- High-Tech Computers (HTC)

Illustrated in Table 3-3, you will find a summary outlining the characteristics of some of the devices currently on the market based on Pocket PC 2002.

Symbol   Symbol Technologies also produces Pocket PC-based PDAs with a built-in barcode reader. Their products have the following options:

- I-Safe PDT 8100/2800 with an integrated 802.11b-compliant wireless adapter.
- PDT 1800 with keyboard.
- PDT 2800, keyboardless (see Figure 3-13).
- PDT 8000 with an integrated keyboard, an 802.11b-compliant wireless adapter, and an optional GSM/GPRS wireless modem (see Figure 3-14).
- PDT 8100 with an integrated keyboard, an 802.11b-compliant wireless adapter, and optional CDMA or GSM/GPRS wireless modem (see Figure 3-15).

Table 3-3. Pocket PC Devices

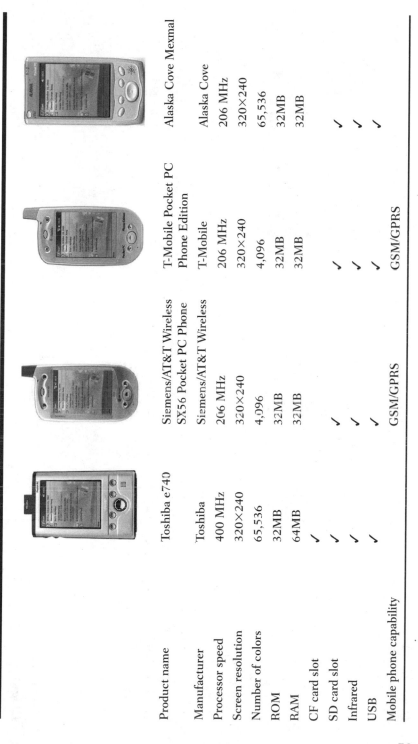

Product name	Toshiba e740	Siemens/AT&T Wireless SX56 Pocket PC Phone	T-Mobile Pocket PC Phone Edition	Alaska Cove Mexmal
Manufacturer	Toshiba	Siemens/AT&T Wireless	T-Mobile	Alaska Cove
Processor speed	400 MHz	206 MHz	206 MHz	206 MHz
Screen resolution	320×240	320×240	320×240	320×240
Number of colors	65,536	4,096	4,096	65,536
ROM	32MB	32MB	32MB	32MB
RAM	64MB	32MB	32MB	32MB
CF card slot	✓			
SD card slot	✓	✓	✓	✓
Infrared	✓	✓	✓	✓
USB	✓	✓	✓	✓
Mobile phone capability		GSM/GPRS	GSM/GPRS	

Table 3-3. Pocket PC Devices (*Continued*)

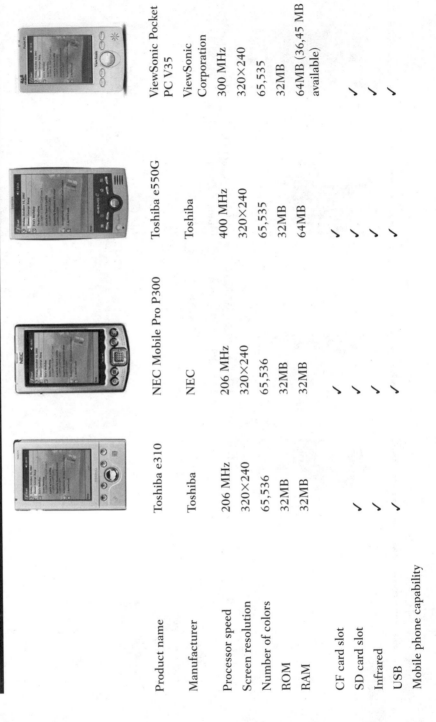

	Toshiba e310	NEC Mobile Pro P300	Toshiba e550G	ViewSonic Pocket PC V35
Product name	Toshiba e310	NEC Mobile Pro P300	Toshiba e550G	ViewSonic Pocket PC V35
Manufacturer	Toshiba	NEC	Toshiba	ViewSonic Corporation
Processor speed	206 MHz	206 MHz	400 MHz	300 MHz
Screen resolution	320×240	320×240	320×240	320×240
Number of colors	65,536	65,536	65,535	65,535
ROM	32MB	32MB	32MB	32MB
RAM	32MB	32MB	64MB	64MB (36,45 MB available)
CF card slot		✓	✓	
SD card slot	✓	✓	✓	✓
Infrared	✓	✓	✓	✓
USB	✓	✓	✓	✓
Mobile phone capability				

	Toshiba e570	HP iPaq H1910	Audiovox Thera	Dell Axim X5
Product name	Toshiba e570	HP iPaq H1910	Audiovox Thera	Dell Axim X5
Manufacturer	Toshiba	Hewlett-Packard	Audiovox Comm.	Corp. Dell
Processor speed	206 MHz	200 MHz	206 MHz	300/400 MHz
Screen resolution	320×240	320×240	320×240	320×240
Number of colors	65,536	65,535	65,536	65,535
ROM	32MB	16MB	32MB	32/48 MB
RAM	64MB	64MB	64MB	32/64 MB
CF card slot	✓			✓
SD card slot	✓	✓	✓	✓
Infrared	✓	✓	✓	✓
USB	✓	✓	✓	✓
Mobile phone capability			CDMA	

Table 3-3. Pocket PC Devices (*Continued*)

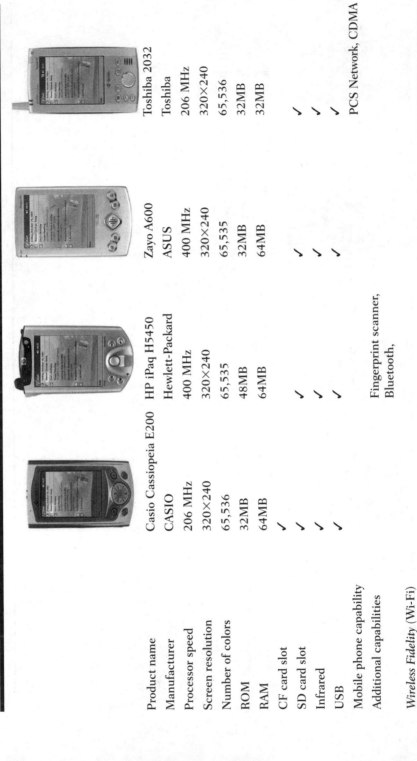

	Casio Cassiopeia E200	HP iPaq H5450	Zayo A600	Toshiba 2032
Product name	Casio Cassiopeia E200	HP iPaq H5450	Zayo A600	Toshiba 2032
Manufacturer	CASIO	Hewlett-Packard	ASUS	Toshiba
Processor speed	206 MHz	400 MHz	400 MHz	206 MHz
Screen resolution	320×240	320×240	320×240	320×240
Number of colors	65,536	65,535	65,535	65,536
ROM	32MB	48MB	32MB	32MB
RAM	64MB	64MB	64MB	32MB
CF card slot	✓			✓
SD card slot	✓	✓	✓	✓
Infrared	✓	✓	✓	✓
USB	✓	✓	✓	
Mobile phone capability				PCS Network, CDMA
Additional capabilities		Fingerprint scanner, Bluetooth,		
Wireless Fidelity (Wi-Fi)				

- *PPT 2800* keyboardless with an 802.11b-compliant wireless adapter, and an optional CDPD or GSM/GPRS wireless modem.
- *SPS 3000* with a jacket for Compaq/HP iPAQ, a barcode reader, and an 802.11b-compliant wireless adapter.

Research In Motion

BlackBerry, like Palm, Inc., created an OS and a device that uses that OS, as illustrated in Figure 3-16. BlackBerry does license out its devices for sale to other vendors, such as Compaq (the new HP) and a variety of wireless phone companies; however, the devices are all the same.

Several flavors of devices are offered by BlackBerry RIM. The smallest device is a smart pager. It offers all the functionality of the larger device, but in the pager size. This also means you get a smaller screen and less area to fit the hardware keyboard, which can be found on every BlackBerry RIM device.

Their newest product is an upgraded version of the full-size device, but with incorporated phone functionality. In this device, you get the always-on features of the standard BlackBerry RIM device, but with the addition of a wireless phone. Table 3-4 provides a broad set of the RIM device options available on the market.

Linux-Based Devices

The only company that is officially mass-producing Linux-based PDAs for the consumer market is Sharp. Their device is on par with the typical Pocket PC device, except for a few factors. It has the advantage of a hideaway hardware keyboard, as illustrated in Figure 3-17, for easy entry of information as well the standard use of stylus-based handwriting recognition.

Even though only one company is commercially producing Linux-based PDAs for the typical consumer, the Linux OS has been adapted by many organizations for devices ranging from a

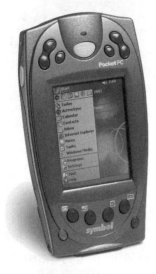

FIGURE 3-13.   PPT 2800.

FIGURE 3-14.   PDT 8000.

FIGURE 3-15.   PDT 8100.

FIGURE 3-16.   RIM device.

Table 3-4. RIM Example

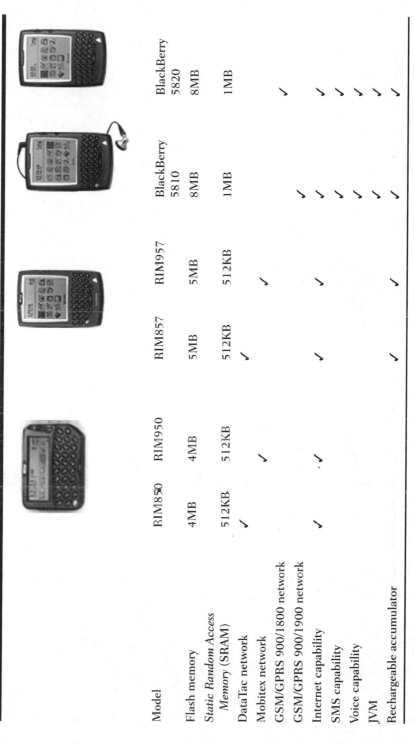

Model	RIM850	RIM950	RIM857	RIM957	BlackBerry 5810	BlackBerry 5820
Flash memory	4MB	4MB	5MB	5MB	8MB	8MB
Static Random Access Memory (SRAM)	512KB	512KB	512KB	512KB	1MB	1MB
DataTac network	✓		✓			
Mobitex network		✓		✓		
GSM/GPRS 900/1800 network					✓	✓
GSM/GPRS 900/1900 network					✓	✓
Internet capability	✓	✓	✓	✓	✓	✓
SMS capability					✓	✓
Voice capability					✓	✓
JVM					✓	✓
Rechargeable accumulator					✓	✓

Table 3-4. RIM Example (*Continued*)

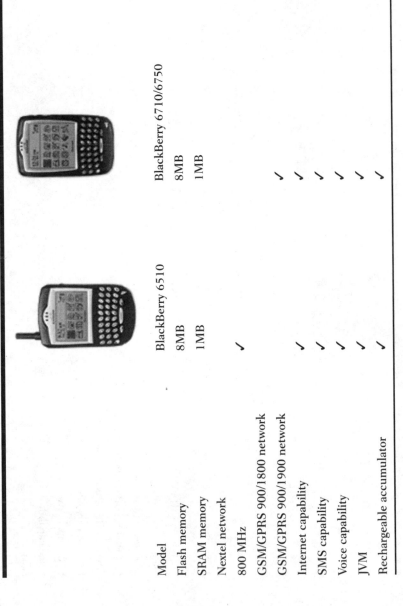

Model	BlackBerry 6510	BlackBerry 6710/6750
Flash memory	8MB	8MB
SRAM memory	1MB	1MB
Nextel network	✓	
800 MHz		
GSM/GPRS 900/1800 network		✓
GSM/GPRS 900/1900 network		✓
Internet capability	✓	✓
SMS capability	✓	✓
Voice capability	✓	✓
JVM	✓	✓
Rechargeable accumulator		✓

Handspring Visor to a custom-made portable barcode reader for use in warehouse inventory processes. The Linux OS can be modified to work on almost any device, and because it is open source, all that is needed is the time, dedication, and skill to configure it properly.

Symbian OS-Based Devices

Currently, Nokia and Sony Ericsson are driving the production of Smartphones running the Symbian OS. Both companies have a variety of devices with more on the way. These devices range from your typical phone to devices that look more like a PDA.

The Nokia 7650 is an example of a typical wireless phone made "smart" by the Symbian OS. It uses the true and tested Nokia form factor while adding a sharp color display and all the functionality of a PDA. Although the Sony Ericsson P800 is an example of what a PDA should look like, this device offers all the functionality of a PDA while being a true Smartphone.

FIGURE 3-17. Linux device.

Some existing Symbyan OS-based devices are as follows:

- *Siemens SX1* (see Figure 3-18)
- *Nokia 7650* (see Figure 3-19)
- *Sony Ericsson P800* (see Figure 3-20)
- *Nokia Communicator 9290* (see Figure 3-21)

Peripheral Software and Hardware

Secondary and tertiary markets have developed in the hardware field as well as in the software area. The following examples of how vendors have used standards illustrate additional areas that must be assessed as devices are deployed or managed in the user community. These examples include SD/MMC card slots, Compact Flash card slots, and HP's jacket concept to extend the capabilities of the devices. An interface and standard for

FIGURE 3-18.   Siemens SX1.   **FIGURE 3-19.**   Nokia 7650.

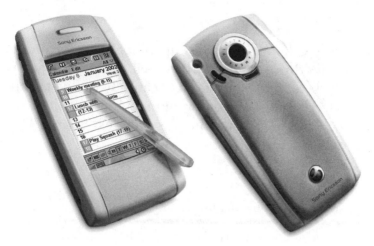

FIGURE 3-20.   SonyEricsson P800.

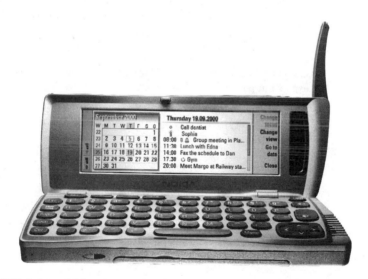

FIGURE 3-21.   Nokia 9290.

providing a keyboard interface not only allows a convenient
method of data entry, it also creates a standard method by which
a brute-force keyboard-based password attack can be mounted
against a physical device. Evaluate the following extensions with
an eye to both increased utility and risk.

HP Jackets

HP iPAQ, formerly Compaq, has the capability to be expanded in a number of ways through an interface that enables the addition of various jackets. For example, you can buy a jacket with an additional battery, additional external memory card slots, or a wireless jacket. Table 3-5 lists some useful jackets from HP and other manufacturers.

Table 3-5. List of HP Jackets by Manufacturer

Manufacturer	Type	Model Compatibility	Description
HP	Compact Flash expansion pack	3600, 3700, 3800 3900, 5400	One slot for CF I, and II card slot
HP	Micro keyboard	3100, 3600, 3700	Small and light QWERTY keyboard
HP	Compact Flash expansion pack plus	3600, 3700, 3800, 3900, 5400	Additional accumulator battery and CF I and II card slot
HP	PC card Flash expansion pack plus	3600, 3700, 3800, 3900, 5400	Additional accumulator battery and PC Card II card slot
HP	Bluetooth wireless expansions pack	3600, 3700, 3800, 3900, 5400	Bluetooth adapter and CF I and II card slot
HP	Dual-slot PC expansion pack	3600, 3700, 3800, 3900, 5400	PC Card II and III y card slot, CF I and II card slot, and additional battery with increased capacity
CoPilot	CoPilot *Global Positioning System* (GPS) jacket	3600, 3700, 3800, 3900	GPS
Symbol	Symbol Wireless *local area network* (LAN) Jacket for Compaq iPaq	3600, 3700, 3800, 3900	802.11b LAN card and additional accumulator battery

Table 3-5.   List of HP Jackets by Manufacturer (*Continued*)

Manufacturer	Type	Model Compatibility	Description
LifeView	LifeView FlyJacket Multimedia Expansion pack	3600, 3700, 3800, 3900	Video input, VGA video output, video-capture capability, built-in CF I and II card slot, and additional accumulator battery

Compact Flash Card Add-ons

If you use a Pocket PC with a *Compact Flash* (CF) card slot (built in or by using a jacket), you can use almost all available CF-slot-compatible hardware to expand your PDA capabilities. Some popular examples are illustrated in Table 3-6.

Secure Digital or Multimedia Card Add-ons

Many modern handhelds include an *Secure Digital* (SD) card slot. Usually, it is used for *SD/Multimedia Memory Cards* (SD/MMC) memory cards, but if the handheld has support for the *Secure Digital Input Output Protocol* (SDIO) protocol, this slot may be used for other expansion devices. Some currently available SD expansions are listed in Table 3-7.

IN CONCLUSION

The preceding inventory of OSs, example applications, manufacturers, and peripherals hopefully has highlighted the great variety of options available to consumers and corporations in

Table 3-6. CF Card Add-ons by Manufacturer

Manufacturer	Type	Model Compatibility	Description
Audiovox	GSM/GPRS WAN card	Pocket PC	Tri-band GSM/GPRS 800/1800 /1900 MHz modem
Blue Monkey	Bluetooth adapter	Pocket PC	Bluetooth card
AnyCom	Bluetooth adapter	Pocket PC	Bluetooth card
Pretec	GPS	Pocket PC	Compact GPS card, CF I
Linksys	Wireless card	Pocket PC	Wireless Compact Flash card, 802.11b-compliant
Symbol	Wireless card	Pocket PC	Wireless Compact Flash card, 802.11b-compliant
Pretec	Wireless card	Pocket PC	Wireless Compact Flash card, 802.11b-compliant
D-Link	Wireless card, DCF-650W		Pocket PC Wireless Compact Flash card, 802.11b-compliant, CF II

the rapidly developing—but still immature—handheld computing field. From a security perspective, the new options and diversity of the environments needing to be controlled all spell greater risks to security.

With an understanding of the handheld computing environment in which we must operate, the next section begins a structured approach to identifying frameworks for exploring an organization's overall PDA security risk. Then we'll define how to develop strong policies to manage that risk, concluding with a case study of a real-world policy enforcement solution at the device level.

Table 3-7. SD/MMC Card Add-ons by Manufacturer

Manufacturer	Type	Model compatibility	Description
Socket	Wireless 802.11b-compliant wireless SD card	Pocket PC devices with SDIO slot, Palm OS 4 and higher	802.11b LAN card
Socket	Bluetooth SD card	Pocket PC devices with SDIO slot	Bluetooth adapter
Toshiba	Bluetooth SD card	Pocket PC devices with SDIO slot	Bluetooth adapter
Palm	Bluetooth SD card	Palm OS	Bluetooth adapter
Spectec	Digital camera	Pocket PC devices with Palm OS	Digital camera with 300K pixels matrix
Veo	Digital camera	Palm OS	Digital camera with 640×480 pixels matrix
Palm	Digital camera	Palm OS	Digital camera with 300K pixels matrix
Margi	Digital projector connectivity	Palm OS	Margi Presenter-To-Go SD card that enables Palm handhelds to connect directly to digital projectors

HANDHELDS IN THE ENTERPRISE

WHEN, WHAT, AND HOW

In the previous section, I introduced the handheld device and how it can be used in an increasing number of roles beyond the standard *personal information manager* (PIM). This section addresses in more detail how handhelds have become a security threat for many organizations, whether they are aware of it or not. In many ways, handhelds are following a path into the Enterprise similar to PCs and related technologies. The primary differences are that Enterprises in a post 9/11 world have become substantially more sensitive to the security implications of the adoption and use of new technology.

When PCs broke into the Enterprise in the 1980s, they were not brought into organizations in a coordinated fashion. Most Enterprises' *information technology* (IT) groups were far too focused on their mainframe-centric organizations to address the new devices. Instead, various departments, groups, or divisions made their own independent decisions on what sort of PC procurements to make. These independent decisions included not only which hardware and software should be purchased and developed, but also how PCs would be integrated with existing business processes. In many cases, device security and, more importantly, data security for the devices were afterthoughts, if considered at all.

It took a long time before Enterprises established or mandated IT departments to centrally plan, procure, and use PCs. However, by the time they began to get control over PCs, *local area networks* (LANs) of PCs had already begun to pop up in various groups, departments, or locations within the Enterprise. Networks tying these PCs together, as well as connecting them with their historically more secure mainframe counterparts, faced an even greater threat and therefore needed to impose centrally controlled policies and procedures. With the advent of networks, it became crucial to control access to PCs and network data in order to protect the Enterprise's proprietary information.

Similar to the evolution and use of PCs and network technology in Enterprises, we are currently at the early stages of IT exerting its widespread influence on the management of handheld devices. At this stage, most Enterprises have not attempted to centrally control the planning, procurement, and use of handheld devices. As a result, most IT departments have not been granted the budget and authority to address handhelds and the potential threats they represent to corporate proprietary data. However, because of today's technology, employees can use their personal handheld devices to transfer, store, and retrieve corporate proprietary data more easily than ever before. Although this can have an incredibly positive impact on the employees' personal and professional productivity, the risk associated with a lost or stolen handheld device is an ever-increasing liability. Further, it is a risk over which most of today's corporations have little to no control, if they are even aware of it.

Chapter 4 looks at how to identify and assess the overall vulnerability of handhelds in an organization. From there, Chapter 5, *"The Components of a Measured IT Response,"* looks at how an organization can generally respond and attempt to control or mitigate the risk imposed by handhelds. Finally, in Chapter 6, *"The How-to Guide: Asynchrony's PDA Defense, Palm's Choice for the Enterprise,"* we will walk through a detailed example of how a current solution can be implemented to execute an Enterprise-level approach to handheld device security.

CHAPTER FOUR

WHEN A HANDHELD BECOMES INFORMATION TECHNOLOGY'S PROBLEM

ASSESSING YOUR CORPORATE RISK PROFILE

Whether or not the IT group, senior management, or employees believe handhelds pose a risk that should be addressed, the reality is that the devices already in place have compromised strategic information assets. In many organizations today, IT has not assumed responsibility for corporate-owned handhelds, even within those organizations that directly purchased *personal digital assistants* (PDAs) for a specific business application or group. More importantly, IT has generally not assumed responsibility for how employees' personal PDAs interact with corporate PDAs.

INFORMATION TECHNOLOGY PROFESSIONALS PROBLEM

The IT professionals driving problem stems from both a security and management challenge. This problem leads to the conclusion that IT professionals must directly set standards and policies, and ultimately control the purchasing of handheld devices. This need to exercise control to address management and security challenges is a reaction to both the rapid proliferation of these devices, and to the increasingly important role they play in employees' information management tasks.

When PCs were introduced to Enterprises, the IT groups of most large organizations were primarily responsible for the management and development of large mainframe systems. Most IT groups, and many organizations for that matter, did not immediately recognize the impact PCs would have on giving individuals control over the automation of business processes and associated information. As a result, IT groups in many cases were not given the authority and budget to purchase and implement solutions using PCs. As discussed previously, they entered the Enterprise in an uncoordinated fashion.

Similar to the early era of PCs, handhelds from various manufacturers are used in Enterprises and run on different *operating system* (OS) platforms. Handhelds, for the most part, are not even coordinated by a department or group; they are still in large part individually purchased devices. Thus, standards for handheld use are rare, and little to no guidance is provided on how they may hook up to the corporate network. As employees try to get their own devices to work, they must each go through the same discovery process or get help from their coworkers, potentially affecting others' ability to get work accomplished. As a result, in addition to the security challenges driving IT management and control, a great deal of lost productivity and other inefficiencies occur.

Security and Management

The theme of security and management as related challenges will emerge in almost all organizations should be addressed as one problem. IT professional's ability to develop a standards-based management of devices within an organization is integrally connected to both the asset management and security challenge.

When PCs originally entered the corporate scene, it caused a tremendous change due to the new hardware and software that had to be managed. To avoid repeated efforts, this management challenge consolidated control into a coordinated and often centralized IT organization.

Handhelds today are in a similar state of change in the hardware and software area, with even more compressed product life cycles than PCs faced in the early 1980s. Without IT management, the rapid introduction of new handheld hardware and software into the Enterprise will not only affect productivity for the individual, but it will also dramatically increase the challenge of providing device management and security.

Enterprises experienced an often painful expansion of PC programs during the early 1980s. Over time, IT professionals were charged with managing the introduction and support of PCs in the organization to help coordinate that growth. Whether Enterprises end up buying handhelds for their employees, or they mingle the use of their personal PDAs with corporate resources, IT professionals must be responsible for the management of handhelds in order to do the following:

- Avoid costly time wasted in getting handheld technology working efficiently with desktops and networks.
- Bring economies of scale in the purchase of handheld hardware and software where applicable.
- Pool efforts to support handhelds in the Enterprise.

Handhelds Touch IT-Managed Assets

As we explore the security issues associated with handhelds, we find that risk primarily stems from the ease with which end-users can integrate devices with IT-managed PDAs that have access to strategic information. In short, PDAs are primarily designed to link to a user's desktop computer and download information. Handhelds can and, for the most part, do interact with many other areas for which IT is responsible. The major areas we'll look at include access to desktops or networks and the information stored on them.

For the purposes of this section, we'll simplify the classification of handhelds into two categories: wireless versus nonwireless. Most normal PDAs do not include wireless capabilities (excluding, for now, expansion cards or slots that make a nonwireless handheld a wireless one). These PDAs can easily be configured and synchronized with an employee's desktop or laptop.

IT employees are generally already responsible for maintaining polices or procedures on how to handle information that is classified as proprietary and may be synced. Most organizations have already established how classified electronic information should be marked and stored, and who has access to different levels of proprietary information. As soon as employees hook up their personal PDA to a desktop and syncs or downloads any information classified as proprietary, this creates a significant vulnerability in an organization's infrastructure. Whatever employees have access to on corporate networks or desktops can be easily downloaded and taken with them on their ultraportable PDAs.

Generally, however, it is easier for IT professionals to control the use of wireless devices, as they often manage standard network transports or "gateways" that have been implemented with security controls already. IT professionals must allow wireless devices to access corporate networks via *virtual private networks* (VPNs) or corporate email systems; employees can generally be prevented from establishing their own VPN or corporate email access using their wireless device. Although the threat is seemingly greater for wireless devices, the more prevalent nonwireless devices pose the greatest risk. Just because of the sheer numbers, the pricing, and the availability of nonwireless devices, they introduce a far greater risk of allowing information to be compromised.

Because IT professionals are already responsible or accountable for desktop access, network access, and system security, they are increasingly compelled to control how PDAs access the existing infrastructure. As soon as an employee synchronizes their PDA and downloads any sort of corporate-associated data, suddenly a hole arises in the security of the Enterprise, for which IT is responsible. IT professionals are also accountable for the risk associated with a PDA that an employee brings in and hooks up to his or her desktop.

Even though IT employees are not formally responsible for the handheld budget, it is something for which they are held accountable. In many cases, handhelds are introduced to the organization by senior management employees, who turn to their IT specialist to help them effectively sync their new gadget with their desktop or laptop. Most IT specialists won't turn down this request, even if they are not formally charged with this type of support. Handheld devices are typically considered to be a companion device that are always used in conjunction with a desktop or laptop. Part of what makes them so appealing is how they can be used to seamlessly keep information in both places up-to-date.

With accountability already implied, IT departments recognize the need to ensure that security and management risks will be properly controlled as handheld devices' role in the organization expands. At this point, the organization must determine the nature of their security risk by focusing on the following questions:

- How are handhelds getting into their organization?
- What kind of information is at risk?
- Are there any projects in the works to deploy mission-critical applications that could expose information on handheld devices?

- Are external factors present, such as federal laws or regulations, that might require extra protection for information residing on handheld devices?

SECURITY RISK MANAGEMENT OF PDAs IN THE ENTERPRISE

What exactly is the risk that PDAs present to the Enterprise? Before you answer that question and start looking for solutions, you must go through a risk-management planning exercise. This exercise will help you assess what is at risk and what needs to be done to monitor and control the risk to your organization.

The following section examines assessing potential risks, discussing the following topics:

- Risk item identification.
- Risk analysis.
- Risk response planning, monitoring, and control.

It seems intuitive that due to the portable nature of PDAs, they can easily be lost or stolen. However, without going through some risk management, one cannot entirely understand how a lost PDA can threaten the Enterprise or its customers.

Risk Item Identification

The first step is to identify who is potentially exposing the Enterprise to risk. In the case of PDAs, the organization should get a handle on how PDAs are entering, what types of employees or groups are using them, and how they are using them. Key questions to study include:

- How are handhelds getting into your Enterprise?
- Are they coming in as personal devices, or are they part of corporate purchases and application deployments?

- What types of employees are using them? What are their roles and responsibilities?

These initial questions should be studied as you formulate strategies to address the risk that handheld devices might pose to your organization.

Risk Analysis

Once your organization understands how handhelds are coming into the Enterprise and who is using them, you can begin studying which type of information is at risk. In most cases, this consists of understanding how the various employees are using handhelds in their ongoing business activities. Is it primarily individuals who have purchased their own PDAs and are using them primarily for PIM applications? Or are groups deploying vertical applications on handhelds for mobile workers?

At the core of your analysis will be a handheld risk classification document, which will be illustrated as we sum up how to assess overall vulnerability. The classification, similar to a data classification exercise, allows an organization to build a matrix including categories such as device types and information assets in order to understand the related risk factors determining an organization's overall vulnerability.

Some additional questions include the following:

- What information are those handheld users carrying on those devices?
- What unintended information are those handheld users carrying on those devices?
- Does your organization have a process in place to classify which information is proprietary and should be actively protected?
- Do employees access any corporate-proprietary information solely on site, or do they have access to it off site?
- Regardless of need, what proprietary information are they accessing off site?

- Do you have plans to deploy mission-critical applications on a handheld platform? If so, do these applications include access or the generation of proprietary data?
- Have you reviewed your business processes recently to identify the impact of handhelds?

Risk Response Planning, Monitoring, and Control

Once an Enterprise has an understanding of how PDAs are entering the organization, who is using them, and what proprietary data is accessed, it can begin looking at risk-response strategies on how to safeguard the information residing on the handheld devices. Some of the questions to address in this process are as follows:

- How tightly does the organization want to control the use of corporate-owned devices?
- What type of policy will the organization adopt for controlling the use of devices owned by individual employees?

Depending on the nature of the proprietary information involved, some organizations may be forced to take a hard-line approach in which no personal handhelds can be used on site or that company-proprietary information cannot be stored on them. Chapter 5 discusses the steps IT groups can take to monitor and control the risk that handhelds present to the organization.

RISK ITEM IDENTIFICATION: WHO IS BRINGING HANDHELDS INTO YOUR ORGANIZATION?

The first step in looking at your Enterprise's risk is to determine exactly how handhelds are entering your organization and where they are coming in contact with your organization's infrastructure. Although a number of companies have begun to implement specific mission-critical applications for handheld devices, most have not actively brought handhelds into their business. However, with at least 20 million devices manufactured and sold by Palm, Inc. alone, a large number of handhelds have been purchased by consumers, many of whom are employees in organizations.

According to the *International Data Corporation* (IDC), roughly 20 percent of purchased devices in 2002 were bought directly by organizations. That number is expected to grow to 29 percent in 2003. Although this is a substantial expansion with more corporate purchasing expected, the vast majority of devices still belong to individuals.[1]

As a result, many of these devices are finding their way into organizations through a "backdoor" without the direct knowledge or sanction of corporate IT departments. Looking at the two major categories of handhelds (wireless and nonwireless), most employees are more than likely using nonwireless devices because of the relatively low cost. Also, most of these devices are being purchased for basic personal information management capabilities; that is, the ability to have schedules, contacts, and notes at their fingertips. For this reason, it is likely that most of these users would attempt to sync their handhelds to their PCs in the workplace.

Far fewer wireless handheld devices enter your organization through the backdoor than nonwireless devices. Unless IT professionals sanction the use of corporate email on wireless devices, most of these devices will touch the corporate network primarily in the same way as the nonwireless devices. Because of the relatively high price point and corporate support required to hook them up to the organization's infrastructure, you will generally be much more aware of the use of wireless devices used.

---

[1]Gaudin, Sharon. "IT Expected to Push Enterprise Handheld Adoption." *Datamation*, 25 June, 2002.

Although a number of organizations are beginning to purchase PDAs either for employees or for specific business areas, the majority of them are personally owned by employees. Most of the Enterprise PDA managers have begun to look at the issue of security. However, many of them are struggling with how to apply policies and impose security requirements on handheld devices that are personal assets of employees.

Although a few organizations are beginning to outfit large groups of employees, such as sales departments, with PDAs, corporate-issue PDAs are the exception today. The majority of PDAs coming into your Enterprise will be through the backdoor, without the support or sanction of corporate IT groups. This backdoor entry of personal devices requires a risk item identification process to focus on who is bringing handhelds into the organization and for what purpose.

Employee Backdoor: The Profile of Employees Who Are Likely to Bring a PDA to the Office

Examining which types of employees are most likely to use a PDA as part of their daily work can be done in a number of ways. We will take three different views of employees:

- *Technogeeks*   Those who gravitate toward new technology on their own.
- *Job function*   Those employees whose personal productivity could benefit from a PDA.
- *Organization or industry function*   What the organization produces or the services it provides.

Technogeeks   A definite set of early consumers exists in almost every organization. They don't all have to be the first to purchase something, but they will often be among the first to use it on a regular basis long before the mainstream consumer. With PDAs, a core set of avid users existed when the first handhelds were commercially available in the mid-1990s. However, within five years, that core set of avid users increased to hundreds of thousands as the number of total PDAs sold reached into the millions by the year 2000.

The number of users and their relative enthusiasm can be gauged by the web sites or communities that support handheld enthusiasts. A brief list of a few of the most common sites is as follows:

- www.brighthand.com
- www.palminfocenter.com
- www.pdageek.com
- www.pdabuzz.com
- www.pdastreet.com

Many of these sites offer news articles on the handheld industry, product reviews, technology conference reports, and so on. Many of them also provide not only links to price comparison and search engines, but they sell the products themselves in their attempt to be a one-stop shop for PDA enthusiasts. One of the most important aspects these sites provide is a forum where enthusiasts can compare notes and offer advice to newbies. One of the most active web sites is bright hand.com[2], where visitors must register with the site in order to post comments or questions with the other 15,000 registered users. They can post to dozens of forums to talk about hardware, software, and industry news, or pose questions. Frequently, many consumers doing research or trying to decide which type of handheld hardware or software to purchase will post questions to get a number of experienced users' views.

The technogeeks are generally comfortable with using technology, especially new technology. They aren't afraid of stumbling through learning how to use the new devices effectively. Unfortunately, this can also affect the organization where the

---

[2]www.alexa.com reports that Brighthand has the largest reach and traffic of all of the other handheld sites. It has 40,000 registered members, 50,000 posts a month in 100 forums, 4–5 million page views, and 200,000 unique visitors from over 120 countries.

individual's productivity might suffer. But once they are up-to-speed with the new technology, they will often become proponents of PDAs, as they will be more than happy to spread the knowledge and provide informal support to other users in the corporation. This has its advantages in that the use of handhelds can increase without any official training or support provided by the Enterprise. Of course, this is also a disadvantage, as the organization does not control how the technology is being implemented, which returns to the potential security threat.

Job Functions   The second category of employee likely to bring in his or her own handheld device is someone whose personal job performance would immediately benefit from the use of a PDA. These types of employees are as follows:

- Executives using PDAs for immediate access to summarized reporting of vital key information.
- Salespeople or mobile workers requiring product data or *customer relationship management* (CRM) system access.
- Account managers and project managers leveraging PDAs for strong project management.

Although these employees are probably motivated to excel in their jobs (which most would be, or they wouldn't be holding their positions for very long), they do not necessarily have to be technogeeks. Of course, the two categories are not mutually exclusive! A number of senior managers and sales executives were the first to purchase PDAs as soon as they were available, because they like to experiment with the latest technology in handheld devices. Although some obvious crossover between these two categories can be noticed, we will assume for now that the majority of employees who have their own PDAs are doing it because they perceive it will help them with their job and career.

For many, the basic PIM functionality is reason enough. Any of the previously mentioned employees can benefit from having all their contacts, their schedule, and other pieces of information at their fingertips at all times. However, many of these employees can also excel by taking advantage of the benefits that wireless handheld devices offer. In fact, it is generally senior management employees who sponsor (or push for) a pilot program for handheld usage, particularly wireless handhelds, within their group. As a result, handhelds that come into the organization from the most visible employee groups are not always via the backdoor. However, as we'll discuss later, this only increases the need for IT professionals to have plans and policies in place to handle the various implementation and security issues.

Organization Type   The third means for bringing PDAs into the Enterprise via the backdoor is related to the function of the organization or group within the Enterprise. One example is a law firm where an individual's time represents the corporate product, and specialized skills must be accessible at any time. For this reason, lawyers, who historically have not adopted technology quickly, have rapidly adopted the BlackBerry devices providing always-on email access. Although we will discuss industry examples at a later point, such as healthcare and patient privacy laws, lawyers exchanging confidential information via email clearly illustrate why IT professionals must understand PDA use to mitigate any substantial risks.

When and Why an Organization Buys Devices for Its Employees

In addition to PDAs entering the Enterprise via the backdoor of personal purchases, certain organizations are purchasing handhelds for specific groups. This is being done for a number of different reasons. Here are some examples:

- CRM applications increase sales information mobility.
- Productivity applications reduce expenses by automating business processes.

- Various other killer applications are being used in categories including executive information reporting, email, and more every day.

In general, the corporate purchasing of devices is driven by applications that offer some organizations a compelling case. However, some sectors that have shown uneven adoption of these applications. Two particular areas or groups in which corporate purchasing has been more widespread includes sales (with CRM-related applications) and executives (with quick access to information and email). From an industry perspective, healthcare, government, and financial services have been rapid adopters, among others.

Healthcare   Numerous examples of handheld deployment can be found in the healthcare field. The portable nature of handhelds, combined with the need for workers to be able to move from patient to patient quickly, makes handheld device deployment a natural fit for numerous applications in the healthcare field.

Let's look at an example from Palm, Inc. The Naval Medical Center in Portsmouth, New Hampshire, is the oldest, continuously running hospital for the U.S. Navy.[3] It delivers state-of-the-art healthcare to naval personnel and their families. Physicians at the Naval Medical Center at Portsmouth needed to find a better way to communicate patient information for personnel on various shifts.

The Naval Medical Center developed and implemented an application running on Palm devices that gives physicians easy access to information from patient files as well as reference information, contact information, and other notes and data. The system replaced handwritten notes on cards, thereby reducing the chance for errors from unreadable notes. It also saved time, allowing for easier and quicker access to information. The

---

[3]Palm Solutions Group. "Palm Success Stories." 2002, http://www.palm.com/enterprise/studies/.

implementation also made the transition between shifts more efficient. Physicians can easily synchronize patient information and case histories so that staff have the information they need literally at their fingertips. In addition to the primary benefits of increasing worker productivity, the organization also benefited from being able to provide a higher quality of service to their patients.

Government   Handhelds have proven to be an effective tool for federal, state, and local government entities. Two examples of PDA implementations from Palm's "Success Stories" include the U.S. Navy and the Alabama Department of Transportation.

The quick and efficient capture of critical data is a must in the U.S. Navy. One example where handhelds greatly facilitate the capture of accurate data is on the flight decks of aircraft carriers. Using Palm devices, Navy personnel aboard the *USS Constellation* and *USS Abraham Lincoln* grade aircraft landings. The previous system required personnel to quickly record landing grades and comments in notebooks. This made accurate record-keeping difficult, and correcting mistakes distracted one from the next landing, particularly when it was hard to see at night.

Two Navy commanders developed a handheld application that enabled personnel to quickly record grades and enter comments for each landing. Errors were reduced and corrected more easily on the backlit PDA devices. The personnel were also able to quickly synchronize the data they captured during their shift, reducing errors from the previous process.

In another example, field inspectors for the Alabama Department of Transportation were using a construction management application to automate the record-keeping processes for their projects. Unfortunately, because the application only ran on a desktop or laptop computer, they were forced to take their notes manually in the field and bring them back to the office to key in later.

The field inspectors and officials from other states asked the manufacturer of the construction management application to create a handheld component that would enable field inspectors to use the software in the field and work in an integrated fashion with their office computers. The handheld application allowed inspectors to download the appropriate information into their handheld, where they could record data throughout their day. This information was synchronized at the end of the shift on the desktop. This made data collection much more accurate and efficient, and it eliminated the inaccuracies of handwritten notes and data entry mistakes. Additionally, the built-in PIM and other productivity tools (such as calculators) in the handhelds reduced the number of other devices that inspectors had to carry with them in the potentially hazardous construction environment.

Financial Services   Handhelds are also utilized by a number of Enterprises to help give them a competitive advantage. An example from Palm's "Success Stories" is Sun Life Financial. Sun Life is an international financial service organization that provides a variety of wealth protection products and services to individuals and corporate customers.

Among Sun Life's products are variable life insurance products. Rapidly changing market information makes staying informed of the latest news a significant challenge. Sun Life wanted to provide its brokers with a mobile solution that could provide immediate and detailed product information that could be updated periodically. This would help their brokers provide more accurate and profitable products and services to their customers.

Sun Life provided Palm devices to a group of their brokers so they could easily access a variety of funds and product information. Updates could be stored and distributed via *Secure Digital* (SD) cards containing quarterly fund performance information. The handhelds provided Sun's brokers with the most up-to-date product information, allowing them to better serve their customers. The PIM applications built into the handhelds also allowed the brokers to have quick and easy access to contact information that could also be kept current with SD cards.

RISK ANALYSIS: WHEN DOES THIS THREAT REQUIRE ACTION?

In the previous section, we looked at identifying handhelds' risks in the Enterprise. Specifically, we discussed how handheld devices enter the organization, who is most likely to use them, and how. The next step in the risk assessment and management process is to gain a better understanding of the impact of risk on the handheld devices. In this section, we will explore:

- What kinds of information can get on PDAs.
- How PDAs interface with various parts of the Enterprise infrastructure.

What Information Is on Those Things?

As discussed previously, the majority of devices within an organization have been purchased by employees who are using primarily nonwireless devices and PIM applications. They want quick access to contact information, calendars, and other information at their fingertips. As a result, a great deal of the information stored on these devices is personal in nature. This could include information such as:

- Personal addresses and phone numbers.
- Financial information such as credit card and bank accounts.
- Web site account and password information.

Personal Information   Although the corporation doesn't have any direct responsibility for employee's personal and financial information, some unintended consequences could occur if that employee loses his or her device. For instance, contact information for the employee's colleagues and coworkers could be stored on the device. If the PDA falls into the hands of a competitor or recruiter, he or she could have instant access to a number of people in your organization if the data on the device is not secure.

Also, just because much of the information an employee keeps on his or her personal device is personal in nature does not mean that *all* of it is. A quick survey will generally uncover the following:

- Business or coworker contact information.
- Customer or partner contact information.
- Company benefits and financial information.
- Internal web site access addresses.
- Business-related web site accounts.
- Business and personal passwords.

The lines between personal and business PIM information can become quickly blurred. The notion of completely segregating personal from business information essentially flies in face of being able to use the PDA and PIM applications effectively. Although it is possible to ask or expect employees to keep two separate PDAs—one for personal and one for business information—in the increasing effort at reducing number of devices (such as phone, pager, and PDA), separate devices may prove to be an impractical solution.

Corporate Information   Most organizations would like their employees to be as effective and productive as possible. Thus, having one device that carries both personal and business-related information seems to make sense. As a result, corporate information will undoubtedly make its way onto the device, whether it is personal or corporate-owned. Some examples of corporate data you wouldn't want to have floating around insecure include the following:

- Customer lists and contact information.
- Direct report personnel information (for managers).

- Strategic partner or vendor contact information.
- Pricing and proposal information.
- Corporate web site account and password information.
- Network or VPN account and password data.

Depending on the responsibilities an employee has, any amount of sensitive information could be stored on a personal device. Thus, it should be an area of concern that this information is being stored, in the majority of cases, on an employee's personal device. As a result, a device being lost or stolen would have a significant impact.

Beyond the PIM use of the device, the increasing array of business applications is driving the purchasing of handheld devices. Applications such as *Executive Information Services* (EIS) and mobility versions of CRM systems are causing the migration of mission-critical data from Enterprise-based computers to handheld devices. This data often has generally understood risks associated with its loss, including operational as well as confidentiality risks.

How Are They Interfacing With My Infrastructure?

In addition to identifying which types of corporate-proprietary information is potentially stored on handheld devices, the organization needs to understand, influence, and, if possible, control who, what, when, and how the data is transferred or synced to the handheld device. Several options are available:

- Direct cable or cradle connections between the handheld device and the employee's corporate laptop or desktop computer.
- Cable or cradle connection between the handheld device and the corporate network, arbitrated by the desktop but managed by a server service.

- Dial-up access between the handheld device and data stored on the corporate network.
- Wireless access between the handheld device and data stored on the corporate network and in email accounts.

Whatever information an employee has access to from his or her desktop or via corporate network connections can be easily synced to a handheld device for quick and easy access.

How to Assess Your Vulnerability

We recommend an exercise to develop a *security risk classification* (SRC) document to evaluate and assess your vulnerabilities. The exercise is based on making assessments of your exposure in the four areas described below. With a focus on the handheld device, this exercise will provide you with a basis and aid in developing a multidimensional approach to assessing your exposure. The SRC is composed of four discreet matrices or tables for each of the following:

- *Industry*   Legal requirements or other industry-specific issues.
- *Organization*   Cultural norms and usage patterns specific to your Enterprise.
- *Access medium*   The range of devices and their capabilities and limitations.
- *Information assets*   Confidentiality, operational risk of loss, and life expectancy.

Each area should be broken down into elements that are specific or relevant to your organization. For example, Table 4-2 displays six different types of Information Assets. Your organization could have 10 more relevant information assets. For each element you can add a row in the matrix or table and assess a value for the Information Classifications in each column. Table 4-1 displays the information classifications and sample values for each classification.

TABLE 4-1. Information Risk Classification Matrix Setup

INFORMATION	CLASSIFICATIONS
Class A	Confidentiality
	100 Highest level of confidentiality
	50 Private and sensitive information about Enterprise/client/vendor
	25 Private information about client/vendor
	1 Publicly accessible
Class B	Operational Risk of Loss
	100 Critical to business operation: Enterprise cannot function if lost
	80 Critical to business operation: Enterprise's ability to operate seriously impaired if lost
	60 Critical to business operation: Client/vendor's ability to operate seriously impaired if lost
	40 Important to business operation: Enterprise/client/vendor negatively affected if lost
	20 General value to Enterprise/client/vendor
	1 Of no material business consequence if lost
Class C	Retention/Access Requirements
	100 Real-time access, no expiration
	80 Real-time access for 9 months, 48-hour retrieval for 7 years
	65 Real-time access for 6 months, 48-hour retrieval for 7 years
	50 Real-time access for 1 day, 48-hour retrieval for 7 years
	30 Real-time access for 9 months, no retrieval after expiration
	20 Real-time access for 6 months, no retrieval after expiration
	10 Real-time access for 3 months, no retrieval after expiration
	1 Real-time access for 1 day, no retrieval after expiration

TABLE 4-2.   Risk Classification Matrix Measuring Six Information Types

Information Assets	Class A	Class B	Class C	Total
Employee contact information	100	40	1	141
Customer contact information	100	80	50	230
Vendor contact information	100	100	20	220
Customer sales data	100	60	20	180
Network passwords	100	100	100	300
Corporate financial data	100	100	50	250

How to Develop the Classification Framework   The set of matrices will be used to set the objective levels of risk that any PDA can be measured against. Once the specific PDA and its associated industry, organization, access mediums, and information assets are defined, the PDA will be assigned an overall risk classification. The numeric score assigned to each PDA will represent its risk across a number of dimensions, or values within the matrix. Security response policies will need to be developed to assign to PDAs based on the resulting numeric score. These policies that are assigned based on the numeric scores will result in some PDAs with relatively slight security measures and others with substantial security precautions.

The overall security exercise can be implemented with a wide spectrum of complexity. It can range from a simple, single-dimensional matrix with a cumulative numeric assessment, as illustrated in Table 4-1, or as an Enterprise-wide multidimensional approach to security policy implementation, requiring multiple matrixes such as those illustrated in Table 4-2.

The following section illustrates how this process, much like the development of a standard data classification document, is dependent on establishing the right values and models for its success. In order to illustrate the use of this framework, Table 4-1 uses the information matrix with the following three key values for evaluation: Operational Risk of Information Loss;

Confidentiality of Information; and Retention Requirements for the Information. In Table 4-1, higher scores representing greater risks are established for each dimension or value. The cumulative score for all values of a particular information asset represents its overall information risk as defined by the matrix.

Once your values are assigned to the matrix and a scoring model has been assigned to each value, each asset can be measured by the matrix. In this case, the assets would include the types of information. The next step requires the development of a comprehensive categorizing of the information assets you expect to be on the PDAs. The following implementation of the information matrix defined in Table 4-1 is illustrated in Table 4-2. Table 4-2 demonstrates the results of an evaluation of six types of information using the matrix.

The score results can be used to assign policies to various PDAs based on their level of risk as defined by the matrix. In this example, the PDAs containing information, which has a score at various numeric thresholds, can be required to implement various policies, the enforcement of which we will illustrate in Chapter 6. In general, a set of increasingly strong security measures can be developed as a response strategy. You can use this framework for scoring and then assign the threat level to a particular user's PDA based on the information assets expected to reside on that particular PDA.

Although we have focused solely on the information assets area in our example matrix (Table 4-2), a complete SRC should also consider the other areas. For instance, access mediums, which refers to the range of ways that handheld devices come in contact with an organization's data, must be included in any Enterprises' assessment. Collectively, we increasingly need to address the network-enabled applications that can run on devices, such as Microsoft's Terminal Services Client on network-aware PocketPC devices. The ability to reset a password or to change user permissions should be factored into your risk assessment and can be included as elements in a matrix addressing the range and degree of vulnerabilities for the access mediums area.

Understanding How to Address Industry-Level Issues Part of assessing your Enterprise's risk is recognizing which internal and external factors affect your business. For certain industries, regulatory considerations must be covered. Some industries, as we will illustrate, have specific requirements to meet regarding information security, while others face a less regulated environment with lower overall risks. However, generally speaking, all businesses face some level of obligation to provide basic levels of protection for their client and employee information.

Healthcare Looking at the healthcare field, the obvious regulations affecting the flow and processing of electronic information is the *Health Insurance Portability and Accountability Act* (HIPAA). HIPAA was passed in 1996 and in August of 2002 it had its final regulations approved, further implementing the act. This law addresses the responsibility of healthcare organizations to safeguard patient information, which they fundamentally do not own but rather hold in trust on behalf of the patient. The law outlines a number of objectives that must be enforced whether the information resides on a mainframe system or on a PDA.

In addition to addressing the responsibility of healthcare organizations to safeguard patient information, HIPAA also requires the US Department of Heath and Human Services (HHS) to develop standards for the maintenance and transmission of information (also referred to as Administrative Simplification). HIPAA is not simply an IT concern, but an Enterprise issue. The protection and security of electronic health information is just one aspect that must be addressed in the context of the entire Enterprise. Looking simply at how handhelds are secured is an important but small piece of the overall puzzle. For instance, securing the data on a handheld device is no good if the processes for how handhelds are used and how confidential information is accessed are not analyzed in relation to the rest of the healthcare organization's processes. As a result, unfortunately, the implementation of PDA security in this context often requires cross-department efforts.

Many Industries Introducing Regulation   Like healthcare, the financial services field currently has to interpret and implement requirements associated with Gramm-Leach-Blily[4]. Even the much broader requirements under the recent Sarbanes-Oxley Act could affect how businesses address the question of PDA security. However, looking beyond the specific industry examples, most Enterprises need to consider their overall exposure to the loss of information or breach of access on handheld devices. As described in the other numerous examples, it is easy to acquire some of this confidential information on handhelds, but it is not so easy to determine exactly what is or isn't secure if the organization has no comprehensive approach or set of policies.

RISK RESPONSE PLANNING AND CONTROL: WHAT IS A MEASURED IT RESPONSE

In preparing for the next chapter, let's take an inventory of our current level of organizational readiness. You have to walk before you can run, and if your organization has only a limited approach to implementing security requirements, you might need to prepare before moving forward.

After determining which kinds of information and functionality constitute a risk, and which external factors would necessitate a comprehensive security approach for handheld devices, the next step is to examine the user access controls and privileges that must be extended to handheld devices. Addressing a measured response to handheld security risk management

---

[4]The GLB Act was signed into law in November, 1999, and repeals the Glass-Steagall Act which prohibited banks, securities firms, and insurance companies from affiliating. One of the key provisions of GLB is the requirement of all financial institutions to disclose to customers their policies and practices for protecting the privacy of non-public personal information.

is the focus of the next chapter, but in preparation review the following questions:

- *Information classification*   Do you understand your information security requirements and what data must be protected?
- *User controls*   Do you have strong user- and group-level controls for roles and privileges?
- *Business process understanding*   Has your Enterprise established a clear process for obtaining information?
- *Overall security requirements*   Does the organization have strong security criteria in place that can or should incorporate the use of handhelds?

If your organization has a strong security approach as part of the general IT network and asset-management areas, many of these questions will already have been answered and a set of handheld policies can be developed. However, if your organization has had a limited approach to security, you find that once you go beyond your user-password policies, you will be working from scratch. At this point, assess your organization's readiness to incorporate handheld security criteria into current security practices.

Information Classification

Below are some examples of questions that you can review for your organization before you begin to formulate and finalize your measured response to handheld device security risk issues.

- Does your organization have defined classes of information that define what is confidential and what is nonconfidential?
- Has your organization established a process to classify information?

- Are there any guidelines to determine how the different classes of information should be handled?
- Who has access to various classes of information?

User Controls

- What kind of information do users have access to?
- How volatile is the data on your user's devices?
- Do users have the right to take information with them when they are not in the workplace?
- Can users identify what information is considered proprietary or confidential, and what they should do to protect it?
- Which user classes exist and how varied are their access rights?

Business Process Understanding

- Which business processes require proprietary information to be accessed outside the office?
- Which level of information volatility or frequency of change do current processes require? For example, do devices contain read-only data or do they update master databases or sources?
- Do automated business processes currently require that information leave the office (either electronically via handheld/laptop computers or on paper)?

Overall Security Requirements

- Do you have written security policies in place?
- Does your organization have security policies for your systems (desktop, laptop, networks, server/mainframes, and so on)?

- Do you have policies defined for how employee access, transmit, or send and receive information?
- Do you perform any audits or checks to see how well to which policies are adhered?
- Do you have any policies established for handheld computers?
- How do you tackle putting controls on devices used by individual employees?

CHAPTER FIVE

THE COMPONENTS OF A MEASURED IT RESPONSE

In the previous chapter, we took a closer look at how hand-helds are entering the corporate world as well as how and why they are a risk. Having established a methodology for under-standing these risks, we will now take a look at what constitutes a measured organizational response to these emerging threats.

If you have not already defined handheld-related policies in place, the first step, in most cases, is to establish and roll out updates to existing security policies to address handheld devices. In many cases where organizations already have well-defined security requirements, policies need to be updated to include handhelds. One of the most difficult areas to address is how to handle the use of employees' personal handheld devices in the workplace. The most security-conscious organizations will not have much difficulty restricting their use, but most organizations will undoubtedly struggle to find both a mea-sured and reasonable level of control for their handheld secu-rity policies.

POLICY MANAGEMENT IN THE ENTERPRISE

First of all, enterprises must review the current general security policies and procedures they have in place. This involves a host of issues that include, but are not limited to, the following topics:

- Data/information classification.
- Access to information/data on the network (including servers, mainframes, and so on).
- Access to information/data on desktops and laptops that connect to the network.
- Access to information/data on standalone desktops and laptops.

In addition, the security policies and procedures must be accompanied by:

- A deployment strategy to roll out the policies and procedures.
- Training programs to ensure that employees are aware of the security policies affecting them.
- Auditing or reviewing how well the security policies and procedures are actually being used.

It is a blessing if you have no security policies in place to deal with a clean sheet and define your policies and procedures with any preconceptions. On the other hand, it might be a curse as you attempt to define and enforce security policies in an area where they have not existed before. Users, by nature, will resist anything that they perceive to get in their way or make things more difficult. Hopefully, we can all agree that handheld devices will only expand in use on the enterprise landscape. As a result, we must now begin incorporating them into our existing and developing security strategy. Whether or not the handheld devices are officially sanctioned, the *information technology* (IT) staff or security groups must devise a strategy for how to address their presence and their potential threat to infrastructure assets.

The Scope of a Good Policy Management Strategy

Overall security strategies must include many factors to provide a comprehensive approach to risk management. For example, data encryption does not add much value without a viable encryption key strategy. If the data is protected but the keys are weak or become compromised, the data is not effectively protected. However, it is also possible that even with the use of encryption and an effective key strategy, an organization may become the victim of a malicious application installed on a device or computer that intercepts a password before it is deleted. In short, a good security policy needs to take a holistic approach to risk management in which handhelds represent only one component.

However, in order to narrow our focus here to the topic of handheld computing, a strong policy should specifically address the following factors with respect to PDAs:

- Software distribution
- Audit and monitoring
- Policy enforcement

Software Distribution   To provide the needed security for employees' handheld devices, the security software must be installed and updated each time the software manufacturer releases software patches and updates. This is essential for eliminating potential backdoors and weaknesses discovered periodically by hackers as a result of exploitation attempts. If the software is not updated and a vulnerability is discovered, a malicious hacker may exploit it to gain unauthorized access to the device's resources.

From a control standpoint, handheld devices ideally contain only applications and data that are required for the expected work of users. Applications and information that are not directly related to the job or organization could represent malicious software such as viruses and Trojan horses that constitute a threat to the organization's security.

Various third-party synchronization solutions can distribute applications and data from a central server to handheld devices across the organization. These solutions can also check for the presence of required software and reinstall it if necessary. Some examples of this third-party software include the following:

- XTNDConnect Server by Extended Systems.
- Scout Sync Server by Syncrologic.
- Afaria by Xcellenet Mobile Communication.

Auditing and Monitoring   Auditing and monitoring the contents of handheld devices is a good policy across the enterprise. This process consists of an IT administrator installing security software onto an employee's handheld device, yet quite often enterprising users find ways around corporate security policies to install their personal preferences and pet applications. Also, the implementation of handheld security policies is almost certain to be resisted by experienced handheld device users who are not accustomed to stringent security requirements. Most likely, a certain percentage of users will attempt to install their own applications, which could cause a potential threat to data integrity and system security.

For those organizations that desire or require a higher level of control over handheld devices in their organization, auditing and even ongoing monitoring of employees' PDAs on a regular basis should be considered. An example of higher-level control offered by XTNDConnect Server includes the capability to monitor handheld device contents, including files and databases. This provides the administrator with the ability to control applications and databases on handheld devices across the organization without having physical access to them.

However, it is important to understand that these solutions are generally agent-based, meaning a software footprint is required to ensure compliance. Although a soft reset by the user will not compromise this agent, a hard reset allows the user to completely delete everything on the device so that it is restored back to its original factory-default configuration. Although a hard reset also removes all the application functionality and information generally intended for protection, it allows the user to bypass any security measures to acquire control over the device. Some sort of audit process will need to be addressed to account for this if a comprehensive asset-management policy is your ultimate goal.

Policy Enforcement   Operational and sustainable policies require both flexible tools and a clear focus on manageability. For example, enforcement means selecting strong tools that enable sufficient control over handheld devices. However, policy enforcement also entails managing the policies themselves, and this, at a process level, can sometimes be overlooked. Before you explore the specific nature of how to control handheld devices (which Chapter 6, *"How-to Guide: Asynchrony's PDA Defense, Palm's Choice for the Enterprise,"* discusses in great detail), a key requirement, regardless of policy, is a mechanism that can update the devices. Depending on the nature of your infrastructure and tools, essentially two key methods can be used for deploying new policies onto a secured device:

- An administrator may manually override security policies through physical access to a device.
- An administrator may deliver updated security policies via a network transport using an agent on the desktop or PC with which it is synchronizing.

If the administrator cannot deliver updated policies in an automated fashion, he or she will have a serious management problem if dealing with a large number of devices. Security policies must be grounded by actionable and sustainable methods based on the tools and resources your organization is prepared to use.

KEY SECURITY REQUIREMENTS OF THE AGENTS DEPLOYED TO HANDHELD DEVICES

When addressing the security of handheld devices, the enterprise must consider a number of areas including:

- *Authentication*   Securing general access to the *operating system* (OS) and the device's application functionality.
- *Encryption*   Securing data via use of encryption, which provides protection that remains independent of the current OS and applications.
- *Fail-safe measures*   Measures in place that will protect the information in the rare or unlikely event that authentication and data encryption fail to protect the information.

Authentication

As discussed earlier, due to the portability of handheld devices, they are prone to being lost or stolen. Compared to the cost of lost hardware, the loss and possible compromise of the applications and information on the device should be a much larger source of concern.

Most handheld device Operating Systems do not have strong authentication and security alternatives built-into the OS. As a result, unless there is an additional security solution installed on the device, the task of taking additional security measures falls on the shoulders of the end-user. In most cases, if left to the end-user, access to the device would be quick and easy without any sort of authentication process. To some degree, the requirement of authentication is directly at odds with the entire notion of what handheld devices provide: easy access to information at the users' fingertips. Any form of authentication slows down or impedes such access.

After working with a number of organizations that have deployed handheld security solutions, I have observed that users often complain about the initial deployment of authentication solutions. Healthcare organizations, particularly physicians, provide a good example of an important user group that has begun to embrace the use of handheld devices. However, physicians in a hospital setting are also a highly vocal and empowered user community. Despite the nature and sensitivity of the information that physicians deal with, and the importance of protecting that information, they still resist the inconvenience of a mandated authentication process. Indeed, organizations must come to terms with how they can best manage the tradeoff between security and alienating users who feel that the authentication process impedes the use of handheld devices.

When looking at device authentication, a number of methods and options are available. Some provide a higher level of protection than others, but a tradeoff exists between cost and the available technology; you must also factor in the ease of use when deciding what sort of authentication to pursue. At a summary level, the main forms of authentication include the following:

- Single-factor authentication:
  - Key code (fixed code such as passwords or *personal ID numbers* [PINs])
  - Challenge-response scheme
  - Biometrics
- Dual-factor authentication:
  - Smart cards
  - Authentication tokens

Single-factor authentication is the most common and widespread form of authentication for not only handheld devices, but also for most systems. It is essentially the use of a single password or passphrase that the user knows and (presumably) does not share with anyone else. Single-factor could also include a username as well as a password. It is, however, limited to either something you know or something you have, but not both, as would be in the case in dual-factor scenarios.

Single-Factor Authentication   Single-factor authentication can be made more effective in numerous ways, all of which will be discussed in following sections:

- Using relatively strong passwords.
- Utilizing safe methods of password entry.
- Applying password management rules and enforcing them.
- Securely storing password information using encryption.

Password/Passphrase Options: Strong Versus Weak Passwords One of the most common threats that users facilitate is weak passwords that are relatively easy to guess or crack. First of all, the password should be comprised of a minimum number of characters. The more characters in a password, the greater the statistical difficulty of identifying the password through various brute-force methods(e.g. guessing the password). Depending on the password's maximum length, the number of possible combinations increases at an exponential rate for each additional character in the password. This implies that as a user is requird to choose a password with an increasing amount of characters, a vast number of additional possibilities are added making access by an unauthorized user increasingly more difficult by use of "brute force" methods. However, even sufficiently long passwords can be considered weak if they are common words, names of family members, and so on. Also, one of the most common mistakes many users make is to use the same password for a number of different accounts.

Additionally, the range of possible characters is another factor that can make the password strong or weak. Including both upper- and lowercase alphabetical characters with numbers and even special characters can make a relatively short password much stronger than a comparatively long password composed entirely of lowercase letters. Finally, the use of multiple occurrences of the same character in a password can make it weak (a1a1a1a1). Although users may not be expected to understand these issues, as we construct a policy around password management, we need to establish and enforce a particular level of strength in our authentication solution.

If words are to be used as passwords or passphrases, they should not be chosen out of common dictionaries. A variety of well-known password-breaking techniques use words from common dictionaries. During a password attack, a code-breaking application uses words from common dictionaries and utilizes different forms of each word (such as all small characters, first character capital, and so on).

However, as we have discussed, when using handheld devices, organizations are faced with the tradeoff between ensuring security by using strong passwords and making the authentication reasonable enough so that it does not detract from the device's usability. Ironically, with many different systems requiring unique passwords and the average professional faced with having to log on to a number of systems on a regular basis, many users of handheld devices have turned to storing these user IDs and passwords on their handheld devices.

This storage of passwords on what probably remains a high-risk device can lead to unintended repercussions caused by the establishment of strong password policies without considering usability. Requiring users to keep strong passwords on their PDAs will discourage them from using them effectively (or cause them to seek ways around the use of a strong password). However, the use of a password or single-factor authentication can be sufficient without requiring 32-character passwords if other methods are deployed, including data encryption and fail-safe protection methods against brute-force cracking. Each enterprise will need to take a look at their user base and decide on the appropriate requirements they need to deploy for handheld device security.

Methods of Password Entry   When looking at single-factor password authentication for handheld devices, other methods are available for entering passwords:

- Soft keyboard.
- Graffiti and other handwriting recognition input methods (such as a transcriber).
- Onscreen buttons/symbols.
- Hardware keys/buttons.

Soft Keyboard and Graffiti   Most handheld devices offer options for inputting characters, such as a soft keyboard that enables the user to tap the corresponding character on the screen. Additionally, many devices offer some form of handwriting recognition system, such as the graffiti offered on Palm OS devices. However, when combined with the use of strong passwords, both of these methods are not always satisfactory for many users who seek quick access to the information on their devices.

Fortunately, a number of security solutions for handheld devices offer alternative password-entry options. These options include the capability to link hardware buttons to specific characters for password entry. This capability can also link onscreen buttons or symbols to specific characters. Thus, the user can create a pattern of hardware buttons or onscreen buttons or symbols to quickly enter a password correctly.

Onscreen Buttons and Hard Buttons   Although onscreen and hardware button alternatives increase the usability of strong passwords, they can also increase the risk of password compromise. Someone could observe the pattern of buttons or onscreen symbols used to gain access to the device. Additionally, the data behind the links must be stored in a secure manner so that it cannot be easily intercepted or read from the device.

Another option that some handheld security solutions surprisingly do not offer is the option to mask the characters, that is, replace them with asterisks or dots as the password is entered. With character masking in place, it is much more difficult for someone to observe a user's password if he or she is standing in close proximity to the user entering a password.

Password-entry methods must be considered in a solution in order to make the use of strong passwords more palatable so that they do not detract too much from the end-user handheld experience. Again, the enterprise must decide where the line should be drawn between creating secure methods and detracting from the usability of the devices.

Password Management   Single-factor is by far the dominant model for handheld authentication in use today. In combination with the data encryption and fail-safe measures discussed later, the single-factor authentication provides a reasonable level of security at a reasonable cost to implement. If you choose to deploy single-factor authentication, the following steps should be taken in the overall password management approach:

- Clearly define and communicate policies.
- Ensure the integrity of password data storage.
- Enforce password syntax.
- Enforce periodic password expiration.
- Include the ability for password recovery.

Challenge-Response Authentication   One variation on the standard approach to the username/password authentication method that we will briefly address is the challenge-response model. This is commonly used in systematic handshakes like that of the popular *Secure Socket Layer* (SSL) method incorporated in *Hypertext Transfer Protocol* (HTTP) browsers.

When the user or system requires authentication from another one, a challenge is sent to the party that requires authentication. A response is then generated based on a shared secret known to both sides. The shared secret may be one piece of information or combination of several pieces of information (an encryption key, algorithm, and so on). The authentication system compares the system's challenge to the user-generated response and calculates what the expected response should have been to the challenge. If results of this comparison are equal, a high probability exists that the authentication system is accurate. However, the possibility of false positives would allow unauthorized access. A small probability exists that the user may provide the correct response without knowledge of the shared secret.

For example, if the challenge is a four-digit *personal identification number* (PIN), the response is a four-digit PIN calculated from the challenge. Each challenge has 10,000 possible results. Thus, each time a probability of 0.0001 or 0.01 percent exists that the response was calculated without actual knowledge of the shared secret. To reduce the possibility of a wrong authorization, the system may require three cycles of challenge-responses. In this case, the total probability of a false positive is 10^{-12} ($0.0001 \times 0.0001 \times 0.0001 = 1 \times 10^{-12}$). This level of accuracy is likely sufficient for most applications.

Dual-Factor Authentication   Dual-factor authentication is the combined use of something an authorized user knows (a password or PIN) and something they have (a smart card or key). Dual authentication brings forth a stronger form of authentication than single-factor because the single password or PIN alone will not gain authorized access to the protected information. Common dual-authentication applications include ATM machines, which require the ATM card and correct entry of the user's PIN. Also, a large number of the more security-conscious enterprises require that employees only gain access to secure areas through an identification card in combination with a PIN or password.

Alternatives such as authentication tokens, biometrics, and smart cards could all be used to compliment a username/password authentication or be used on their own as a single-factor solution. The following examples provide a framework for understanding the wide range of options available as you consider extending the single-factor authentication model.

Authentication Tokens   RSA Data Security offers an example of dual authentication. One of their solutions offers two-way synchronization using a dynamic SecureID code generated each time the user needs access to the system. Users have a special device whose function is to generate an ID code that is used along with a known PIN (or password). The authentication server generates this code and compares it with the code provided by the user. If the codes are equal, the user has the appropriate ID-generation device and the system can perform the second-stage verification against the PIN or password.

RSA Data Security also offers another interesting authentication method. The second code is delivered to the user via a mobile phone when he or she logs into the system. The user receives the code via voice or a *Standard Management System* (SMS) and enters it after the appropriate system prompt. The cell phone guarantees that the code is delivered to the correct person. Also, if the cell phone is stolen with the password, this method will not be compromised by an unauthorized user because it also requires the PIN, assuming the user did not type it into his or her phone.

However, dual-factor authentication is not very widespread or easily available. Part of the reason for this is that although the technology is available, the cost of applying these solutions can frequently add a significant amount to the handheld solution, such as 50 percent or more. Once again, the organization must determine the tradeoff between opting for the higher level of authentication versus the added cost of doing so.

Interestingly, some security scenarios being explored by organizations include using the handheld itself as the second factor in the authentication. In this method, an embedded certificate on the device authenticates a user to a remote system. However, this does not help us with our problem and rather only increases the importance of controlling use of the handheld.

For our challenge, a related and better example includes the use of a smart card reader incorporated into the PDA, allowing a separate physical smart card to be a necessary component of authentication. PITech has one such solution based on *radio frequency* (RF) transmissions to a customized case on a particular line of Palm OS devices. Certain manufacturers produce specialized smart card readers for handheld devices, and smart card readers may be used with PDAs that have the *Personal Computer Memory Card International Association* (PCMCIA) expansion slot.

Axcess Mobile Communications, Inc. produces several BlueJacket expansion jackets for Compaq iPAQs with integrated smart card readers, as illustrated in Figure 5-1.

Biometrics   Dual-factor solutions are increasingly utilizing a major and emerging component known as biometrics, which confirm an individual's identity based on a measurable physiological or behavioral factor. Examples of behavioral-based biometrics include speech patterns and handwritten signature analysis. Physiological-based biometric measures are typically physical attributes such as fingerprint checks, iris scanning, and so on. Although someone can use a password in single-factor authentication or have possession of a card and know the required password in dual-factor authentication, a biometric factor is not something that can easily be faked or simulated.

Technology has developed to varying degrees for many different biometrics. However, although some may be capable of implementation using today's technology, their accuracy rate for false positives (allowing an unauthorized user access) and false negatives (disallowing the valid user) may be unacceptable. Additionally, the "Big Brother" stigma is present here, and some people may feel uncomfortable about having one of their per-

FIGURE 5-1.   Compaq iPAQ with Bluejacket expansion pack.

sonal characteristics stored in an online database for authentication purposes.

If the technology provides an acceptable level of false positives or negatives for your policies, this may be worth considering. Perhaps the largest roadblock to the implementation of biometrics is that of cost. Certainly, a number of options are available for implementing biometrics on handheld devices, but most of them are either proprietary and have been designed for a specific application or have a significant cost. The question becomes evaluating the marginal benefit of the biometric or other approaches in relation to the increased cost.

Some examples of PDAs on the market with built-in biometric authentications include the HP iPAQ 5455 (PocketPC 2002) and the CLA Secure PDA with embedded Linux (see Figures 5-2 and 5-3).

Smart Cards   Smart cards are used widely in various areas such as banking (debit cards and vault cards), access cards, and multifunctional cards. A smart card is generally a plastic card with a small chip placed under a pad with contact points, and it sometimes contains a power supply or clock generator.

FIGURE 5-2.   HP iPAQ 5455.

FIGURE 5-3.   CLA Secure PDA.

Smart cards solve a significant issue: the user's difficulty remembering long passwords. Smart cards also implement a more robust authentication scheme using a certificate-based infrastructure or a *public key infrastructure* (PKI). The simplest and cheapest smart cards contain long password or authentication keys. More powerful smart cards can perform a challenge-response authentication scheme with an authentication server using certificates via a wireless medium.

Dual- Versus Single-Factor Authentication   Now that we've discussed various authentication measures from simple passwords to biometrics, how does this apply to handheld devices in your organization? Although it is apparent that single-factor passwords are the weakest form of authentication, a balance must be obtained between ease of use, cost, and security. Although dual-factor and biometric authentication solutions are available for handheld devices, they are expensive, and the technology has yet to be completely reliable. Thus, you can adopt certain strategies to make single-factor authentication more secure. Also, when used in combination with security strategies in addition to authentication (e.g., data encryption), single-factor authentication can provide a sufficient authentication solution for handheld devices.

Data Encryption

Data encryption provides a strong secondary level of protection above and beyond the authentication measures discussed previously. Encryption provides data protection in the event that the device OS is compromised through measures such as the physical removal of memory chips for direct access by chip readers. The following points should be carefully reviewed when determining your encryption-related policies:

- Encryption algorithm.
- Encryption algorithm key length.

- Tradeoff performance versus security.
- Control and recovery of keys.

The security policy will identify the data to encrypt, evaluating the state and use of the information, and determine the specific key length and encryption algorithm to be used.

Encryption Algorithm and Key Length   The encryption algorithm and key length determine the degree of risk that the encrypted data may be accessed without knowledge of the actual key. The key length available depends on the particular encryption algorithm (see Table 5-1).

Algorithms, unlike keys, should be publicly known and well tested. Because algorithms represent the complex mathematics used in conjunction with a key to create encrypted data, new or proprietary algorithms tend to have a higher risk of vulnerability. Although they may benefit from a security perspective because of their relative obscurity, they will generally weaken your overall solution. Don't build your own. However, each well-known encryption algorithm has its own limitations and should therefore be carefully selected.

Data Usage: In-Transit, Temporarily on, and Retained-on Devices   The challenge of encrypting data beyond the algorithm and key entails addressing the various states of information as they are transported to and from the device. Data

TABLE 5-1.   Encryption Algorithm and Key Length Table

ALGORITHM	POSSIBLE KEY LENGTH (BITS)
DES	56
Triple DES	168
AES	128 (standard), up to 256 (nonstandard)
CAST	128

may require encryption in transit to and from the device, and once it arrives on the device, it may have different requirements, depending on its intended use. Data created on or brought to the device may require retention or only be designed to exist for the length of its use within an application. Here are some examples:

- Data intended to be stored on the PDA, such as email synchronized periodically from the desktop or network using *Post Office Protocol* 3 (POP3) and the *Simple Mail Transfer Protocol* (SMTP).

- Data not intended to be stored on the PDA, but rather accessed when needed from the organization's network, such as email provided in a web-centric model via *Internet Message Access Protocol 4* (IMAP4) or access to *Hypertext Markup Language* (HTML) or *Extensible Markup Language* (XML) documents.

Each of these models will have different requirements for encryption designed into the tools you select for protecting on-device information. Although your encrypting agent on the device will address this during its implementation, the following provides a little additional detail regarding the various methods of data exchange and storage.

Data Delivered to or from the Device via a Trusted Transport Data may be delivered to or from the PDA in a variety of ways, but generally when the transmission is within a limited physical area, it tends to be trusted or highly probable that there is little or no possibility for interception. Generally, these data exchanges are synchronization activities via the following primary transport mediums:

- *Universal Serial Bus* (USB) or RS-232 cable.
- Wireless, close-range Infrared transmission (*Infrared Data Association* [IrDA]).

These synchronization methods offer varying levels of inherent security. Some risk is involved, however, and it may be possible to intercept synchronized data, including the user or system password transferred between the desktop and the PDA.

USB or RS-232 cable synchronization is the most secure because no way exists for intercepting data, except when connected physically to the cable, which is highly unlikely because the cable is not long and can be easily watched.

IrDA is also relatively secure due to its limited range. Although it's possible that someone could intercept infrared emanations through another nearby receiver, the limited physical range of this action makes this scenario unlikely. However, some security solutions provide the capability to disable the use of IrDA ports to ensure that this cannot take place, especially as newer PDAs provide increased transmission power and a line-of-site range for infrared use.

Data Delivered to or from the Device via an Untrusted Transport   When data moves over *local area networks* (LANs) and *wide area networks* (WANs) en route to the PDA, chances are the PDA is using a public network for its transport medium as well as standard RF-based technologies, including the following:

- Bluetooth LAN connection.
- *Wireless Fidelity* (Wi-Fi) through 802.11x standards.
- Direct network connection (a LAN card or modem connection).

Bluetooth and Wi-Fi connections are inherently the least secure because they transmit RF-based signals over wider distances, generally without any default encryption. This allows other devices within a reasonable proximity or along the network path traveled by the information to readily intercept data. The Wi-Fi connection may have *Wired Equivalent Privacy* (WEP) encryption enabled, but it has vulnerabilities that make protection less efficient. Bluetooth is more vulnerable than Wi-Fi because no encryption is available by default, and only device authentication is provided. High-level tunneling protocols that provide security with encryption, such as a *virtual private network* (VPN), should be used in these cases, but standard synchronization, such as HotSync (for the Palm OS platform) or ActiveSync (for PocketPC platforms), doesn't provide for the standard use of protected protocols.

As in all security measures, addressing not only the handheld, but the greater context in which the security solution must exist, remains important. In the case of wireless networks based on Bluetooth and Wi-Fi adapters, because they may work at a range of dozens or even hundreds of meters, it is highly desirable to operate them within a relatively secure area to minimize unauthorized traffic interception and sniffing. The reality is that airwaves are difficult to control, and encryption provides the best standard approach to mitigating the risk of this information being intercepted in transit to or from the PDA over an untrusted transport.

In general, the best approaches come back to the use of secure network connections via SSL, VPN, and other related applications. In this case, data is protected between the handheld and the server, generally incorporating a certificate-based challenge-response layer to authenticate the session.

Data Stored on the Device   We will spend much more time in Chapter 6 addressing data stored on devices, but here we pause to examine temporary data, which is not intended to persist on the device after its use. Data known to persist poses a clearer requirement. When data is temporary by design, the expected retention and use of the data prior to its deletion must be understood. This often requires understanding the specific application in question, as illustrated by the following email-related examples.

The *Research In Motion* (RIM) device, known for its always-on email, provides a wireless email capability protected in current releases using the *Triple DES* (3DES) encryption algorithm. This capability enables the user to delete each email after viewing, followed by a step that prevents the device from retaining any email fragments in memory.

Another example is the use of a PocketPC-based device with a VPN connection to the enterprise server via a public network. If the user deletes each email after viewing, the information is expected to be destroyed permanently. These application assertions should always be tested.

In the following chapter, we discuss the array of options for encrypting data retained on the PDA in order to protect the data in the event the device is lost or stolen. Data protection schemes must be used to protect any vital data stored on the device.

Threats to Data in Memory   A valid data protection scheme includes authenticating users in order to restrict access to unauthorized individuals. However, although user authentication prevents most unauthorized attempts to access information, it is still possible for unauthorized users to gain access to data under certain circumstances. Generally, OSs prevent most users from accessing system resources and files, but an individual with the appropriate training, experience, and resources can read files and data by removing the hard disk or memory chip from the device and accessing files through a chip reader. This case provides a framework for you to assess which level of protection is good enough for your organization. The following sections provide a slightly more detailed discussion of evaluating threats.

Bypassing Handheld Device Authentication: Performing a Memory Dump   The majority of PDAs store user and system information in the device's *random access memory* (RAM) or Flash *read-only memory* (ROM). When access to information stored on a device cannot be obtained through accessing the file directory and structure, the low-level access may be used to retrieve data from the files stored on the device. Generally, a memory dump is a technique that can be used to get raw data from the device; however, significant additional analysis would be required to recover high-level files. As a result, file or object contents are available immediately. A number of handheld devices (Palm OS, PocketPC, and RIM) have a manufacturer debug mode available that enables a developer to do a complete memory dump and retrieve it from the device. In some cases, the manufacturer's debug mode is hotwired in the device's OS and can be triggered regardless of whether or not the device is locked.

Physical Disassembly of the Device   If the debug mode is unavailable, it is possible that the data from the device can still be retrieved. The chip itself can be removed from the handheld device. If the handheld device's memory is nonvolatile (such as Flash RAM), special actions are not immediately required to preserve data on the chip after removal from the device. Thus, it is possible that information may be retrieved and recovered from the memory chip.

If the RAM chip is volatile, more complex actions are required to preserve the chip contents. A dynamic memory chip requires a power supply and special handling in order to keep the internal memory cells intact during removal, but theoretically, it is possible.

For example, RIM OS-based handheld devices have Flash RAM storage, so it is possible to remove the memory chip from the device for investigation. Also, these devices don't have the capability to do hard resets to remove data permanently from the handheld memory.

Although it is highly unlikely that the data on your organization's handheld devices is at risk of physical chip removal, it is still possible, and you need to assess whether this should be factored into your risk planning.

Device Bit Wiping

One fail-safe method for protecting confidential information on handheld devices is bit wiping. In this method, sensitive or confidential information is deleted when it is not needed or if intrusion is detected. This keeps information out of the hands of unauthorized users.

Discouraging Brute-Force Attempts   The first thing an intruder will attempt in an effort to compromise a locked device is a brute-force attack against the password. This involves systematically running password combinations against the device. If the handheld device enables unlimited attempts at entering the correct password, it is highly likely that this type of attempt will succeed over some period of time. Brute-force password attacks can be defeated by a number of countermeasures, including the following:

- Forcing a time delay between password-entering attempts. This ensures that the intruder cannot enumerate passwords quickly.
- Increasing the time delay after each unsuccessful attempt.
- Limiting the number of password attempts prior to a permanent failure. After the password-attempt limit is exceeded, the following actions may take place:
  - The device may require that it be unlocked by a special administrator key.
  - The device may initiate a data deletion or wiping to avoid any unauthorized access.

Ensuring Data Is Truly Deleted   A problem often faced by IT and security professionals is determining the point at which information has truly been deleted from electronic media. Extreme policies generally end up physically destroying the media, akin to shredding a document beyond reconstruction. However, short of destruction, a number of options are available, from a simple delete based on the OS's basic features to a multipass bit overwrite of the media. A policy must be implemented to clearly define what it means to delete data.

Data Existence After a Simple File Deletion   It is a well-known fact that when a user deletes a file or record in most OSs, all the data still resides in memory storage. In fact, most systems have a readily available "undelete" functionality. Beyond this feature, major OS providers often offer third-party-developed applications that specialize in data recovery. These are popular applications because users occasionally delete data accidentally, creating a valuable function for these software products in the nonhacker world.

When a user deletes a file on most systems, the OS simply deletes the reference to the storage area where the information is actually stored. An appropriate analogy is to think of a library where the librarian uses a card catalog to find the location of a book in "storage" on a shelf. When the librarian needs to remove or replace a book, he or she first locates the card in the catalog to determine the book's location. When he or she removes the index card, the book has not actually been removed from storage until the librarian removes or replaces the actual book on the shelf. The image of a library and changing cards can be extended to how information is stored and deleted on most computer OSs. When a user deletes a file, he or she is simply removing the reference or "pointer" to where the data is actually stored.

Handheld devices are not different in this respect. When a deletion is executed, the OS only deletes the index pointer for the file to improve the overall performance. The data is only deleted when the same memory area is overwritten by new data.

An additional consideration is the use of built-in Flash memory (many PocketPC devices have built-in Flash-based storage). When writing data to Flash memory, each block (usually 64KB) must be erased first. The erase operation is relatively long, and the memory manager attempts to perform the erasing. If you write a memory block into the Flash several times, the Flash storage manager may write several memory blocks in a series, instead of overwriting only one memory block in an effort to improve performance. All this leads toward more instances of your data and not fewer.

Overwriting Each Byte in Memory   Given these facts on how most OSs and devices store data, some utilities are available to completely delete data securely. However, no easy way currently exists for doing so on most handheld devices, yet some security products perform a secure data wipe under certain circumstances in order to protect data from unauthorized access, as will be illustrated in Chapter 6.

A standard, secure data deletion process is necessary for handheld devices for two reasons:

- A user might want to ensure that sensitive information is completely erased from his or her device. For instance, in some government facilities, the general policy is to make sure that no confidential or sensitive information can leave the facility, which includes handheld devices. In this case, the user may wipe data in the RAM storage and external memory cards.

· In the event that a handheld device falls into the wrong hands, the user would want to ensure that the data is completely deleted to avoid any unauthorized access. In this case, the device must be equipped with appropriate third-party security software containing data-wiping capabilities. Data should be able to be deleted after detecting a brute-force password attack or after exceeding a specified time.

What Else: Antivirus Protection, VPNs, and So On

The preceding discussion about authentication, encryption, and data wiping covers the core areas that must be addressed to ensure end-device security policies can be enforced. The following sections address other widely discussed security-related areas that should also be considered as a part of your overall policy.

Virus Threats Another factor to be considered in a comprehensive handheld device security policy is the threat from viruses. They are a great risk to data integrity because many of them corrupt data and can cause software malfunctions. What is a virus? Generally, a virus is a small application that always has following characteristics:

· The virus application will attempt to make copies of itself. These copies may be transferred to another computer either via infected software or a network.

· A virus does not require user permission to execute, and it has a mechanism to be launched automatically by the OS or existing application software.

A virus may have additional characteristics:

- A small footprint size, taking little space in memory.
- The capability to access the user's address book to send copies of itself to new targets.
- Stealth, encryption, and polymorphism. Certain viruses make their detection more difficult. These viruses use a stealth technique to mask their presence in the system. They may also encrypt and modify their own code to make their detection by an antivirus system less efficient.
- Data corruption. Many existing viruses can perform destructive actions under certain circumstances. For example, some may corrupt data accidentally due to bad code; however, others have been created specifically to destroy a user's data and applications.

The vast majority of viruses have been created for Windows-based machines with the Intel x86 processor families, and a number of viruses act on UNIX-based computers and other OS platforms as well. Actually, very few viruses have been spread to handheld devices, and thus the virus threat for PDAs is low at this time. However, due to the increasing popularity of PDAs and the creation of network-enabled devices and Smartphones, the virus threat will increase. Antivirus software will become essential for handheld devices, similar to the need for antivirus software for desktop computers.

The following software manufacturers provide antivirus programs for handheld devices:

- Symantec Corporation, *www.symantec.com*
- Kaspersky Lab, *www.avp.com*

Secure Networking and VPNs   Network-enabled PDAs are convenient because they have remote access to an organization's network. Employees may send and receive corporate email, use a web browser to get data from Enterprise web servers, or have direct access to the Enterprise network via *Transmission Control Protocol/Internet Protocol* (TCP/IP) public networks. Many handheld devices support external telephone modems or wireless *Code Division Multiple Access/ Global System for Mobile Communications/General Packet Radio Service* (CDMA/GSM/GPRS) modems, and they allow the use of WANs. Also, a number of PDAs with built-in or add-on capabilities for Wi-Fi and WAN wireless adapters are available (refer to Chapter 3, *"The Power Resource Guide to Understanding Where Security Must Be Achieved,"* for examples).

However, this convenience also comes with an increased threat. The increasing use of public WANs has another side: the possibility of data intrusions and the threat to overall Enterprise network security. To minimize security risks, handheld device users should utilize a secure network connection to access the Enterprise network remotely.

Handheld device users may also employ a different set of secure network connections depending on their needs, such as SSL, *Secure Shell* (SSH), and VPNs. Some handheld devices may have built-in secure network connection support (see Table 5-2).

TABLE 5-2   Secure Networking Capability by OS

	PalmOS	PocketPC	RIM OS
SSL	Third-party solutions	Built-in	Built-in
SSH	Third-party solutions	Third-party solutions	Third-party solutions
VPN	Third-party solutions	Built in for PocketPC 2002	Third-party party solutions

QUICK ASSESSMENT: WHICH LEVEL OF CONTROL ARE YOU PREPARED TO ACHIEVE?

The Enterprise must first assess the desired level of control and certain questions must be answered:

- Which features will the organization allow end-users to control?
- Which features will the organization preset and control for end-users?
- Which features will the organization prevent end-users from accessing?

Controlling Device Purchasing

An Enterprise security policy should describe which handheld devices can be used inside the organization. These policies range from forbidding the use of any handheld devices to wide-open approval of employees using their personal PDAs. For example, some time ago the U.S. State Department prohibited Colin Powell[1] from using a Palm OS handheld device after the decision was made that handheld devices were not secure enough.

   Between the two extremes are a number of security options. In order to gauge how extensive your security must be, you may want to start out with some broad-based positions, such as the following:

- Users may use any type of handheld device.
- Users may use any handheld device, but they must be equipped or have a built-in security solution that meets the organization's security requirements.

---

[1]http://www.businessweek.com/bwdaily/dnflash/mar2001/nf2001038_563.htm

- Users must use only those handheld devices certified by the organization's security group and with the proper ready-to-install security software.

- Users must use only the handhelds purchased by the organization with preinstalled security software and certified by organization's security group.

The first policy option opens up the organization to the most risk. Allowing the users to use any type of device without restrictions makes it difficult to impose any sort of control. General support would become challenging for IT employees, and the security administrator would not know which types of handheld devices are being used and/or which type of security software is being used, if any.

The most security-conscious companies will likely choose the last option where employees only use handhelds purchased by the organization. Additionally, these handhelds will be equipped with security software that is managed remotely. Even with this implementation, the organization should perform audits on the regular basis to ensure that employees use only intended handheld devices with the appropriate security software active.

If the organization does not purchase handheld devices for employees but allows the use of PDAs, the organization's security policy should ensure that handheld devices are installed with security software, and that security policies are enforced. Regular audits should also be performed. Handheld devices without security software must not be used for keeping confidential and important Enterprise information. Any unprotected device with confidential information represents a risk the organization can avoid with the implementation and enforcement of appropriate security policies.

Specify the Manufacturer, OS, Software, and Peripherals

The organization's security policy may describe the type of handheld devices that can be used and which peripherals may be used in conjunction with those devices. Peripherals include accessories such as memory storage cards, wireless modems, etc. The control of peripherals used is important because the security risk may be different for the same handheld device if accompanied with different peripherals. For instance, a hand-held device with a wireless modem that allows access to a corporate network can pose a much higher risk than a handheld without any peripherals. Security policies may further define the handheld device characteristics, such as the following:

- Manufacturer
- OS
- Security and other software or applications
- Peripheral devices

The handheld device manufacturer determines which additional components and built-in software applications can be installed on the device. For example, some handheld devices can have built-in security-related features such as a smart card reader, a fingerprint scanner, and so on. However, most devices available on the market do not currently have such security devices.

The handheld device's OS is one of the most important PDA characteristics. It determines handheld vulnerabilities and security features available to the end-user without the addition of any third-party software. These characteristics are dependent on the OS version as well. For example, Palm OS 3.X versions utilize a vulnerable system password. Also, the OS debugger software is hotwired, so it can act as a backdoor on a locked device. In comparison to the 3.5 version, Palm OS v4.0 stores the system password in a more secure manner, and the OS debugger software backdoor is disabled on a locked device. Thus, some handheld device OSs should be considered more secure than others.

Although the OS may be considered secure if no backdoors have been verified, the built-in security software may still be considered weak. A good example is older PocketPC devices. The device itself generally does not have any backdoors (Compaq's iPAQ is an exception; see Chapter 8, *"White Hat Hacking Threats and Mitigations"*), but the built-in security software is weak. The password can only be a four-digit PIN, and the total number of available password combinations is 10,000, which can be attempted in a relatively short time.

Handheld device peripherals create additional security issues. If a nonconnected handheld device is considered secure, that does not imply that the same handheld device with network connectivity has the same level of security. If a handheld device has wireless capabilities that enable it to connect to both a public network and the Enterprise network simultaneously, the device may inadvertently create an unprotected gateway into the Enterprise network from the public one. A good example is the Hewlett-Packard iPAQ 5450. It is equipped with a built in Wi-Fi adapter that can connect to an Enterprise network. The 5450 also has a built-in Bluetooth adapter that can connect to the Internet via a mobile service, if available. In this case, the employee may ignore the Enterprise network security, work with Enterprise data, and have an uncontrolled public network connection at the same time.

Limit Nonintended Uses: Restricting Applications

Management or control over software applications for handheld devices is an important part of establishing a security policy. Although third-party security solutions may resolve any limitations of a weak default solution, third-party software can also create new backdoors that are difficult to close. If users are allowed to install unapproved or untested software, the organization faces a risk that the device may be compromised. As always, IT must ensure that a balanced approach is maintained to safeguard not only the device, but also the end-user's need to install everything from utility programs to games to horoscope applications. In the end, however, security administrators

TABLE 5-3. Checklist

	RECOMMENDATION	IMPORTANCE
1.	Develop a security policy to solve handheld security issues regarding PocketPC, Palm OS, RIM OS, Symbian OS, or others.	Must have
2.	Train handheld users in security awareness and risk management.	Must have
3.	Upgrade handheld firmware (ROM) with the latest software updates to ensure security patches can be applied immediately.	Must have
4.	Establish a process, preferably automated, for performing security checks and audits.	Must have
5.	Use physical access controls, such as photo IDs, smart card readers, and so on, when employees enter building and other security areas.	Must have
6.	Take a complete inventory of all used handhelds.	Must have
7.	Turn off wireless access points when they are not used.	Must have
8.	Apply a security policy on the handheld after a hard reset.	Must have
9.	Enable all security features on the handhelds.	Must have
10.	Use data encryption on the locked device.	Must have
11.	Set up maximum encryption key lengths.	Must have
12.	Deploy IPSec-based VPN technology for wireless communications.	Good to have
13.	Test and deploy software patches and upgrades on a regular basis.	Must have
14.	Require handhelds to have strong passwords (user and administrator passwords).	Must have
15.	Change all passwords regularly.	Must have
16.	Enable user authentication for handheld device management.	Must have
17.	Ensure that management traffic uses wired LANs only.	Good to have
18.	Deploy security products that offer enhanced cryptography and authentication.	Good to have

TABLE 5-3.   Checklist (*Continued*)

	RECOMMENDATION	IMPORTANCE
19.	Fully understand the impact of deploying any security feature or product prior to deployment.	Must have
20.	Track handheld security products and standards (such as *Federal Information Processing Standards Publications* [FIPS] and others), as well as threats and vulnerabilities of emerging handheld technology.	Good to have

should consider restricting the use of nonintended software applications on handheld devices. In order to ensure conformance, regular audits or inspections of handheld devices should also be performed.

Checklist

Table 5-3 illustrates a checklist to help you identify what level of security and control makes sense within your organization.

THE HOW-TO GUIDE

ASYNCHRONY'S PDA DEFENSE, PALM'S CHOICE FOR THE ENTERPRISE

In the previous chapters of this section, we looked at how handheld devices are entering the corporate world and how they can pose a risk to any type of Enterprise. Once an organization has decided to address such a risk, it is generally left up to the *information technology* (IT) support organization to formulate a security plan. This involves a host of issues, including establishing standards and policies for handheld device usage in the organization. These policies must be defined, communicated, enforced, and audited to be successful.

Once the Enterprise has decided to adopt *personal digital assistants* (PDAs) within its organization, it must tackle some difficult policy decisions and identify the tools for implementing those policies. At this point, the organization must decide whether or not to allow employees' personal handheld devices to interface directly with corporate-owned desktops. Perhaps equally difficult is how to enforce such policies. The previous chapter discussed considerations such as user authentication, encryption, and other security aspects that would help the IT staff put together an enforceable response to the risk of handheld devices. Once the organization has defined its handheld device policy, a specific solution can be implemented across the Enterprise to enforce those policies, and this chapter provides an example of such a solution.

Handheld devices, which began primarily as an end-consumer product, have now grown and adapted more toward the Enterprise market. Recognizing the importance of the Enterprise field, Palm, Inc. has put a significant amount of effort into making sure future product offerings fill the needs of organizations looking to roll out handheld implementations.

ASYNCHRONY'S PDA DEFENSE: A BRIEF HISTORY

As in most PDA-oriented solutions, the story starts with Palm. In the effort to market products to the corporate world, it quickly became apparent to Palm and other handheld device manufacturers that security had to be addressed. With a growing number of devices coming into organizations via individual or group purchases for specific applications, many handheld manufacturers were running into sales challenges around the issue of security. Many organizations, recognizing that no control existed over the ability to secure PDAs once deployed, were holding off on large Enterprise buys until a solution could be addressed.

Additionally, some security-conscious customers, such as various organizations within the U.S. government, actually enforced a policy that any handheld device leaving a secure area would have to be destroyed. The underlying concern was that even if handheld devices were reset, the classified information stored in memory would not be deleted. This issue was obviously a stumbling block for handheld manufacturers selling devices in large numbers to security-conscious organizations.

PDA Defense, a division of Asynchrony Solutions, was brought in to help a federal government organization address the concern of ensuring that classified information was effectively erased. Asynchrony's first customers requested that the bit-wiping routine, used as a fail-safe measure, be invoked voluntarily and wipe the memory in a specific pattern. So, similar to how bit-wiping applications in PCs carefully overwrite each memory bit in a specific pattern, PDA Defense developed a bit-wiping routine to ensure every bit is overwritten.

In early 2002, about six months after Asynchrony Solutions released their first version of PDA Defense Enterprise for Palm *operating system* (OS) devices, Palm approached Asynchrony looking for an IT-configurable security solution for corporate Enterprises. The majority of handheld device security solutions available at that time were completely configurable by the end-user and did not allow for administrative control over the security settings. PDA Defense Enterprise was one of the first solutions that provided organizations with the ability to customize their deployment of endpoint PDA security.

PDA Defense was initially created by Asynchrony Software's collaborative developer community (*www.asynchrony.com*). One member of the developer community, while using his Palm OS device on an airplane, realized the potentially devastating consequences of losing an unsecured device. Having recently read about the Asynchrony developer community, he decided to locate others with the appropriate skills to create a solution. After some initial investigation, he discovered that the built-in handheld OS security was not strong enough or easy enough for most end-users to use. At the time, password-hacking programs were available for Palm OS devices that essentially deleted the system password on the device. Additionally, after briefly surveying the product marketplace for Palm OS security applications, an opportunity arose for another Palm OS third-party product that offered stronger options for protecting data on the device.

Using the resources available in the Asynchrony developer community, a team with experience in handheld OS programming, security practices, and encryption was formed. PDA Defense's predecessor, PDABomb, was initially released for beta testing to the Asynchrony developer community in October 2000. After three months of intensive beta testing and feedback from numerous members of the Asynchrony community and marketing staff, PDABomb was officially released as a shareware product in January 2001.

The shareware product was targeted primarily at individual end-users concerned about ensuring security on their devices; however, within weeks of PDABomb's initial release, Asynchrony Solutions began receiving feedback from Enterprises that were looking for a solution that allowed an administrator to configure options for all his or her end-users. In October 2001, PDA Defense Enterprise for Palm OS handheld devices was released. This was the first-generation PDA Defense Enterprise product that enabled an IT administrator to preset and lock the security options using PDA Defense for his or her Enterprise end-users.

In the summer of 2002, Asynchrony Solutions released the first version of PDA Defense Enterprise with multiple handheld platforms supported. PDA Defense offered the first product that provided the capability to generate security policy files for Palm OS, PocketPC, and *Research In Motion* (RIM) devices using a single user interface.

PDA Defense provides the powerful capability of flexibly configuring end-device security across multiple handheld platforms within an Enterprise. IT administrators can control as much as they want or allow end-users the ability to control many security options and features themselves. All of this depends on the security policies that the organization sets for different user groups. What sold Asynchrony Solutions' partners and Enterprise customers on PDA Defense was its powerful capability to customize the software deployment for end-users. In the next sections, we'll use PDA Defense as an example for what you should expect in an Enterprise PDA security solution, and we will walk you through the features and configuration options you should expect from your solution.

Deploying PDA Defense Enterprise to an organization consists of four main steps:

1. Create template policy files.
2. Create custom policy files.

3. Deploy PDA Defense and the custom policy files to end-users.

4. Deploy policy file updates as needed.

As the PDA Defense policy files are created, the primary protection features of PDA Defense must be determined and customized by the organization's IT administrator, including:

- Secure password protection.
- Encryption of the data stored on the device.
- Fail-safe bit-wiping measures to ensure the data is not compromised.

Additionally, other features provide additional levels of protection for data stored on the device, such as the following security options:

- Allowing only administrator-approved applications to run on devices.
- Password-protecting applications before programs can be run or launched.
- Blocking functions from end-user access, such as the *infrared* (IR) port.

CONFIGURING AN ENTERPRISE POLICY

Numerous considerations must be made in selecting and implementing an endpoint security solution for handheld devices. Once the organization has determined which policies will be implemented and what level of control the handheld administrator will deploy for endpoint security, the next step is to configure the policies using the PDA Defense Policy Editor. We will cover all the configuration options in a step-by-step fashion, as well as some of the primary protection alternatives available.

After installing and launching the PDA Defense Policy Editor on a standard workstation, the administrator will be automatically prompted to create a new group and enter an administrator password specific to that group. The groups can be viewed and selected on the left portion of the Policy Editor, while the detailed policy information associated with each group can be viewed or edited on the right side, as illustrated in Figure 6-1.

The groups are an important feature in a handheld security solution. Group management allows the IT administrator to create different sets of policy files for different groups of users. For instance, an organization will want to have one set of policies for a group that may not require many restrictions (such as IT), but apply more control over device usage for a different group

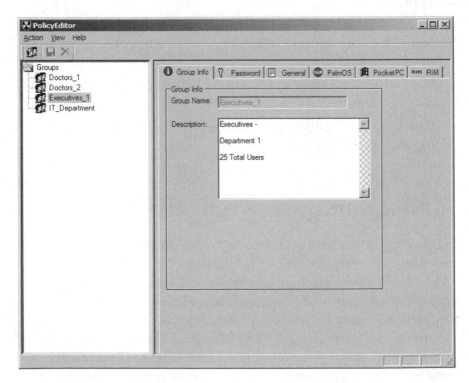

FIGURE 6-1.   Policy Editor with sample groups Doctors_1, Doctors_2, Executives_1, IT_Department, and the group information tab.

of end-users (such as office staff). This is often the case in many organizations.

For example, many healthcare organizations are resistant to making too many restrictions on their doctors. Some doctors even balk at the use of passwords, because some feel that password entry detracts significantly from the handheld experience. Although recent *Health Insurance Portability and Accountability Act* (HIPAA) regulations have increased the need to enforce certain user and application restrictions, healthcare organizations have already realized the need to apply controls over various types of hospital staff. Part of these differences might be organizational (for instance, doctors are not always employees of a hospital, such as hospital staff).

In the end it is up to the organization to decide how much control they should exert over various user groups. As you will observe in the example of PDA Defense to follow, policies can be managed at a group level rather than requiring user-level controls that would generally involve direct integration with Active Directory or *Lightweight Directory Access Protocol* (LDAP) systems to manage user-level permissions. Group-level policies create the best approach to rolling out centralized policy management settings to user communities.

Creating the Template Policy File: Copying a Set of Baseline Applications and Databases

The first step in configuring an Enterprise policy for your organization is to determine the set of applications and databases that you plan to protect via PDA Defense. By default, handheld devices come installed with *personal information manager* (PIM) and other applications from the device provider.

However, if your organization uses any third-party or custom applications that get loaded onto your devices, you will need to create a baseline or template policy file from which you can select your third-party or custom applications for management. This is a critical and important distinction in case your organization has deployed a custom application you want to ensure is selected for encryption or fail-safe bit wiping. For instance, a hospital might need to ensure that a custom medical application deployed on hospital staff's handheld devices is selected for encryption. Yet they don't necessarily want all databases and applications encrypted on their staff's devices, as it might adversely affect the performance and usability of the handheld devices. Thus, the capability to create these organization-specific or custom templates for generating policy files is an important feature that adds to the flexibility of rolling out a handheld security solution.

Figures 6-2 and 6-3 illustrate the encryption and bit-wiping selection windows before and after a template policy file for a Palm OS device is loaded into the Policy Editor. Figure 6-4 shows what the encryption selection screen looks like for PocketPC devices.

Creating the template files is relatively straightforward, but the administrator must start with a device that has the expected configuration, or the applications and databases for each user group that requires its own security policy. Once the baseline set of applications has been determined for each group, the administrator loads PDA Defense Enterprise on the device and

FIGURE 6-2.   Protected database window before importing template.

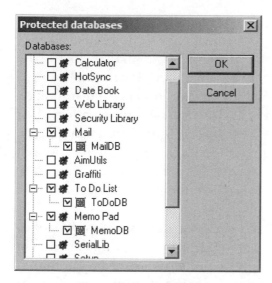

FIGURE 6-3.   Protected database window after importing a template.

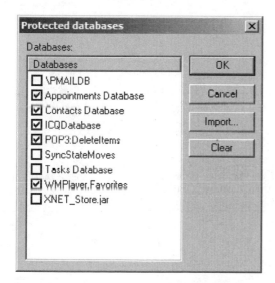

FIGURE 6-4.   Encrypted database window for PocketPC with the PocketPC hooked up to show third-party applications.

runs through a brief series of steps to save the template. After syncing the device, the policy template is saved to the desktop, and the administrator is able to import the template into the appropriate group in the Policy Editor.

If you are setting up PocketPC devices, the process is a little simpler, requiring only that a device with the required set of applications or databases be available and hooked up to the desktop where the Policy Editor is installed. Once this is done, the administrator can simply import the information into each of the areas required. For PocketPC devices, information must be imported in order to configure encrypted databases, encrypted folders, protected applications, wiped databases, and wiped folders, as illustrated in Figure 6-5. Additionally, the process for selecting databases to be encrypted on the PocketPC is illustrated in Figure 6-4.

Once the administrator has successfully imported the policy file templates, he or she can specify protection alternatives for any custom or third-party applications deployed on the handheld devices within the Enterprise. With this information, he or

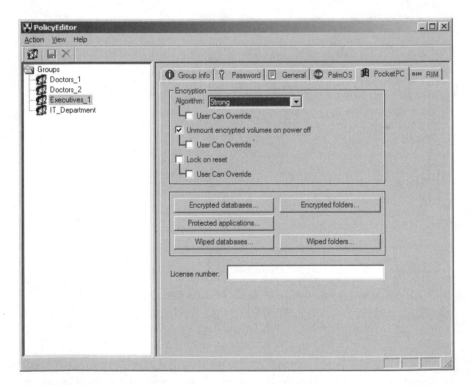

FIGURE 6-5. Tab illustrating options for setting up restrictions on PocketPC devices.

she can begin creating a custom policy file for the organization. Next, we'll go over the authentication policies.

Customizing Policy Files: Allowing User Control over Some Features

A powerful endpoint security solution for handheld devices would allow a large amount of flexibility when it comes to implementing specific features or options. Different organizations, such as healthcare groups, financial services, or government organizations, all have different needs and requirements. Additionally, organizations may want to exert more control over certain users. For instance, a government organization would want to maintain tight control over what users can do with PDAs in a secure and classified area, as opposed to end-users in a non-classified area.

One of the areas the administrator must tackle is whether or not to allow some user control over setting or changing the security features on his or her own device. PDA Defense Enterprise provides the capability to centralize some policy controls while still decentralizing other controls or options. As an example, Figure 6-6 illustrates the "General" settings tab within the PDA Defense Policy Editor.

Not only can the administrator preset whether an option is enabled and what the setting is, but he or she can also specify whether the end-user can modify that setting for many of the features. In Figure 6-6, the "User Can Override" box designates whether the end-user can change the option once that policy file has been deployed.

For instance, in Figure 6-6, the "Auto lock on power off" box is enabled for all users with this policy, and the end-users will not be able to disable this option. The option "If off more than" is also enabled and set to one minute. However, the end-user does have permission to either disable this option or keep it enabled and change the value of the time delay. With the exception of encryption, most of the features can be configured to enable or disable end-user change permissions. We will point these out as each feature is described.

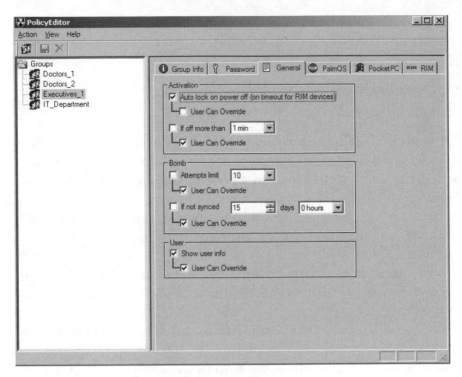

FIGURE 6-6. General tab with the "User Can Override" box unchecked under Auto lock but checked for "If off more than 1 min."

Customizing Policy Files: Determining Authentication Policies

Ensuring well-managed authentication is the heart of the PDA Defense solution. The other key components, including encryption and bit wiping, largely provide additional measures to ensure that authentication policies are enforced under extreme circumstances. Although authentication largely comes down to a strong password, the following range of policies provides the framework within which detailed policies concerning authentication can be managed by IT administrators for their overall PDA community.

Administrator Password   The first step when creating a set of policies for a new group is to determine the administrator password. The PDA Defense Policy Editor will prompt the administrator to set an administrative password when a new group is created. It is the responsibility of the administrator to ensure that he or she creates a strong administrative password and that he or she tracks the administrative passwords associated with each group. The administrator password is the key to allowing policy file updates to be deployed to all end-users in a distributed fashion. If someone tries to send out a substitute policy file, it will be rejected and not applied by PDA Defense unless it is properly authenticated with the correct administrator password. From a business perspective, it is important to protect the ability to make policy file updates. This administrator password feature provides the organization with the capability to:

· Protect themselves from end-users wanting to change the mandatory settings by creating their own policy files.
· Protect their end-users from attempts to change policy files either locally for a single device or globally for an entire group of devices.

In the event the IT administrator needs to deploy an updated policy file, he or she has the opportunity to change the administrator password. Again, the original administrator password must be included in the updated policy file, but it is entered as the old password. The new administrator password is entered in the first password field. In all cases (new and old passwords), the administrator password must be entered twice for verification purposes. Figure 6-7 demonstrates the PDA Defense Policy Editor with the Password tab selected so you can see the administrator password fields.

End-User Password Rules   Figure 6-7 also displays all the options for setting the end-user's password rules. An administrator has the freedom to set up his or her password with no restrictions, whereas users must obey certain password rules:

· A password must meet a minimum length.
· A password must be a combination of letters and numbers.

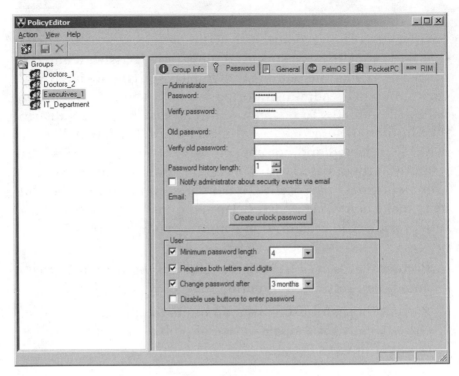

FIGURE 6-7.   Password tab with administrator password fields filled in.

- The end-user must periodically change his or her password.
- An end-user must be able to remember to track end-user password history.

Minimum Password Length   The administrator is responsible for specifying the minimum length in characters for the end-user's password. The PDA Defense Policy Editor allows a set of flexible choices, ranging from 4 to 15 characters. Although it seems self-evident that a reasonably strong password would require as many characters as possible to prevent brute-force attacks, a surprising number of end-users just use a bare minimum of one or even two characters, or even no password at all if given the chance. Again, the organization and administrator are faced with the tradeoff between security for their handheld devices and ease of use. Although a password of any length will always detract from the user experience of a handheld device, the organization has a fiduciary responsibility and must take steps to protect its assets by setting and using password rules.

Use of Letters and Digits   Requiring the use of digits and letters provides another way of ensuring end-users utilize a strong password for handheld authentication purposes. This simply increases the amount of possibilities a password can employ. The more possibilities, the more difficult it is for a brute-force effort to break the password. Also, requiring the use of digits prevents the end-user from using common words or names. More advanced password rules will be implemented in future releases that will analyze the password chosen and reject common words or patterns of characters.

Periodic Password Changes   The longer a user keeps the same password, the greater the risk of device-level authentication being compromised. As stated, requiring periodical password changes provides another way to ensure the end-user utilizes a strong password. The administrator can determine how frequently the changes must occur, which can range from every 15 days to once a year. Upon expiration of the password, when the user unlocks the device, he or she will be prompted to change the password. Ensuring that the user doesn't reuse the same password increases the strength of the password.

Periodic History   When the user is mandated to change a password, or changes it voluntarily, he or she should be forced to use a password he or she has never used before. Otherwise, forcing the user to change it in the first place doesn't add much value to the strength of the password. Many authentication systems have the capability to remember previous passwords. In the PDA Defense Policy Editor, the administrator can select the number of previous passwords the system can remember. The end result is a policy that the administrator can tailor for their organization.

Autolock Features   One of the inherent weaknesses in many of the default security applications is the inability to automatically lock the device when it is turned off. For instance, the early versions of the Palm OS devices required the end-user to perform a task in order for the device to lock. Only the most diligent user concerned about security would bother to take this extra action. Making a security application relevant means that the security features must be either required or easily invoked; otherwise, they simply will not be used by the end user. In the case of PDA authentication, this involves how the device is locked after it is either turned off or not used for a given period of time.

Overriding Autolock (General tab, Palm OS Options screen, PocketPC screen)   One of the most common choices in deploying PDA security is establishing autolock as a constant feature, so that any time the device is turned off, it also locks. Figure 6-6 shows how to force the autolock feature in the PDA Defense Policy Editor so that it is always on for end-users that receive the appropriate policy files. As a result, when end-users view the PDA Defense options, as illustrated in Figure 6-8, while they can attempt to tap and clear the box by "Lock on power-off" to disable automatic locking after power-off, they will not be able to change it."

Time Delay Before Locking (PalmOS Options screen, PPC Screen)   An additional feature corresponding with autolock is the capability to allow a time delay between when the device is turned off, either by timeout or by manually turning it off, and before the PDA Defense locking mechanism is engaged. This is another feature that helps make security more user-friendly and nonintrusive. This feature is commonly employed if the device is used frequently during the day, such as in meetings. With most devices set to timeout every two to five minutes to conserve battery life, it may be desirable to quickly start using the device without having to enter the password. If the time delay is set to 10 minutes, frequent use during the day or while in a meeting might not be affected. However, if the device is left unused for a period of 10 minutes or more, the device would automatically lock, securing the device without any action on the part of the end-user.

As illustrated in Figures 6-6 and 6-8, the administrator can allow the end-user to control this setting while forcing the autolock feature to occur on power-off. Figure 6-9 shows what the autolock and time delay settings look like for a PocketPC device.

PDADefense Options

Activation:
- ☑ Lock on power-off   ☐ Smart
- ☑ If off more than ▼ 1 min

Bomb:
- ☐ Attempts limit
- ☐ If not synced

Owner:
- ☑ Show owner information
- ☐ Stealth mode

(OK) (Buttons...)

FIGURE 6-8. Options screen in PDA Defense showing autolock as enabled and time delay enabled with the same setting shown in the Policy Editor screen in Figure 6-6.

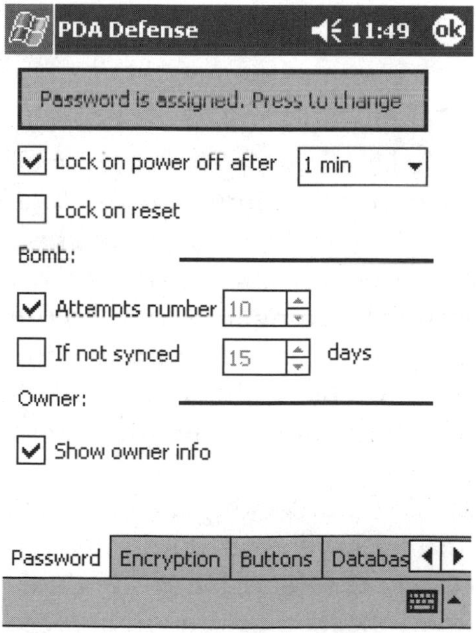

PDA Defense   ◀€ 11:49   ok

Password is assigned. Press to change

- ☑ Lock on power off after │ 1 min   ▼
- ☐ Lock on reset

Bomb:
- ☑ Attempts number │10 ↕│
- ☐ If not synced   │15 ↕│ days

Owner:
- ☑ Show owner info

| Password | Encryption | Buttons | Databas | ◀ │ ▶ |

⌨ ▲

FIGURE 6-9. PocketPC tab showing the same autolock and time delay settings as in Figure 6-6.

The settings in these figures set up a time delay of one minute, and the end-user can change this value or disable it. Although this provides a great deal of flexibility in order to tailor the security settings for individual groups, the administrator setting the policies must ensure that he or she is satisfied with the end policies that will be deployed to various user groups within the organization. Some organizations may want to enforce a specific time-delay period they feel is acceptable, whereas others may want to provide users with greater flexibility, enabling end-users to set any delay period they desire.

Other Authentication-Related Features PDA Defense Enterprise provides a number of other authentication-related features that enable flexible deployments, satisfying the varying requirements of a number of different Enterprises. These additional authentication-related features include the following:

- Temporary user passwords.
- Hardware buttons or onscreen buttons for easier password entry.
- Future incorporation of dual-factor and biometric authentication techniques.

Temporary Password   The single largest support issue associated with deploying handheld security solutions is users accidentally locking themselves out of their devices. As much as administrators hope users will responsibly manage their passwords, the reality is that if they aren't writing it everywhere for others to see, they are most likely forgetting it. At any given time, a certain percentage of users will be unable to enter their password correctly to access the information on their handheld device.

Unless an alternative is available for unlocking the handheld device, the end-user must perform a hard reset that, in effect, deletes the data and applications on the device. He or she would then be forced to rely on a previous backup or sync in order to restore the data and applications. Although restoring a device from a previous sync can offset the worst implications of a lost or hard-reset device, many users will lose a substantial amount of data that was not up-to-date on the desktop since the last sync or backup of the devices. Therefore, without a temporary password capability, preferably one that can be issued remotely, users will lose data.

The administrator can resolve this situation by generating a temporary password that unlocks each user's handheld device. The process of generating the temporary password includes a challenge-response model in PDA Defense and follows these steps:

1. The user generates a unique ID for his or her device.
2. The user contacts the administrator and uses the organization's unique authentication process to ensure that he or she is a valid member of the organization.
3. The administrator uses the unique ID provided to generate a temporary unlock password that is valid for a fixed period of time specified by the administrator.

PDA Defense Enterprise provides the capability to generate a unique device ID from the lockout screen. As displayed in Figure 6-10, the end user clicks the ID button to generate the ID.

FIGURE 6-10.   Lockout screen with ID button and a window with the unique ID example.

This unique ID is based on a unique device identification as well as information identifying the PDA Defense Policy File installed on the device. When the administrator uses the PDA Defense Policy Editor to generate the temporary password, he or she must open the appropriate group that created the policy file for that particular device. In order to generate the temporary password properly, the Policy Editor and the installed policies on the device must share a common administrator password. Figure 6-11 shows how the administrator must enter the device ID and the start date/time for the temporary password.

The temporary password generated by the administrator is valid for up to 60 minutes from the date and time used in the Policy Editor. The administrator must ensure that he or she uses the appropriate date and time for the end-user's handheld

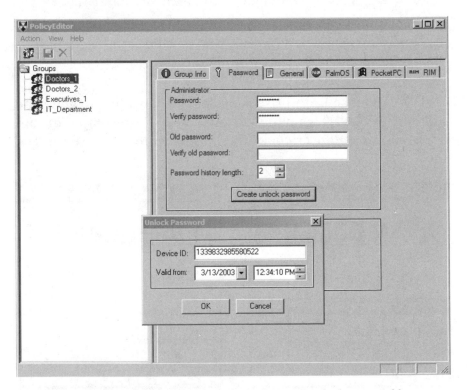

FIGURE 6-11. The "Password" tab with the "Create" unlock password button selected and a unique device ID entered in the "Unlock Password" dialog box including the resulting temporary password.

device. If the administrator generates a temporary password valid for a future time period, he or she must specify the time window in which the temporary password will be valid on the end-user's handheld device.

Once the end-user receives the temporary password and correctly enters it, he or she will be immediately prompted to change the user password at that time. This ensures that the end-user can immediately create a new password, in effect resetting his or her password. If the user happens to select a password that has been used previously, as dictated by the password history, the user will be instructed to select another valid password. Of course, the other password rules will be enforced as well.

Use of Hardware Buttons or Onscreen Buttons   As has been previously discussed, a tradeoff exists between strong authentication and ease of use or handheld device usability. Much of what handhelds have come to represent for users is the quick access to information at the user's fingertips and the ability to record data easily. As a result, certain users consider the use of a strong password to be a distraction from the expected handheld user experience. Again, the assumption is that the organization has determined that the risk or exposure of data loss dictates a security policy that detracts from ease of use. Therefore, authentication, while detracting from the end-user's handheld device experience, is necessary to ensure the integrity of proprietary corporate information stored on these devices.

PDA Defense has incorporated some options to solve the conflict between easily accessing information and the ability to safeguard that information through stronger authentication. PDA Defense provides the capability to link or map characters of a user password to the hardware buttons of a device, assuming that policies allow such features. This enables the end-user to unlock the device by pressing the correct pattern of hardware buttons at the system lockout screen. In the future, PDA Defense Enterprise will offer alternative shortcuts for password authentication using a series of onscreen symbols or buttons. Figure 6-12 shows how the end-user can link characters in their password to hardware buttons available on a Palm OS device.

FIGURE 6-12.   Setting up the password buttons.

However, because mapping password characters to hardware buttons may be considered a weak method of password security, PDA Defense Enterprise also offers the administrator the option of disabling this feature. If the administrator checks the appropriate box to disable the use of the hardware buttons, the end-user will be prevented from using this feature, as illustrated in Figure 6-13. This feature continues to support the philosophy of PDA Defense to provide flexible features that can be customized to fit a wide variety of security needs.

Future of Dual-Factor or Biometric Authentication Techniques
The options provided by dual-factor and biometric authentication will continue to grow, as discussed previously. PDA Defense Enterprise will be enhanced and updated appropriately as technology becomes more widely available at a lower unit cost, or as the market demands it.

Dual-factor authentication will most likely require that custom-designed devices, built to accommodate this kind of authentication, may use PDA Defense Enterprise via a smart card or a similar device. Although a few dual-factor systems are available on the market today, they are not widespread at this time. Dual-factor authentication options available today

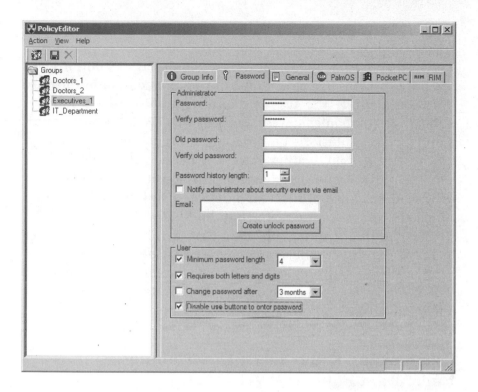

FIGURE 6-13. Password tab highlighting the "Disable use buttons to enter password" checkbox.

generally only support one or two specific devices or are designed for custom-built handheld devices. Although dual-factor authentication provides a much stronger defense against unauthorized access, single-factor authentication using a strong password will be sufficient for most of today's organizations, given the level of protection they require for their devices.

The same is true for biometric authentication. Although it provides a much stronger authentication method, it is too costly and could be considered overkill for most of today's needs. However, it will develop quickly. Several examples of biometric authentication are already used by handheld devices today. Some of the most costly solutions include biometrics based on physiological attributes, such as fingerprints. Unfortunately, most of these solutions depend on customized hardware and software, so most of the major handheld device manufacturers do not offer many solutions "out of the box."

However, some behavioral-based biometric solutions are becoming widely available. Some of the most common software solutions include the use of digital signature recognition. A number of companies have patents in this area, and some end-point security solutions have deployed digital signature recognition. Unfortunately, they have not been widely adopted yet because of the higher incidence of false positives and negatives they produce. Given the tension between single authentication solutions and compromising handheld devices' ease of use, stronger authentication solutions that involve a much higher incidence of discrepancies will not be as widespread. For organizations looking to employ much stronger authentication at the risk of detracting from ease of use, dual-authentication options may be usable.

Customizing Policy Files: Determining Encryption Policies

The second major level of protection for handheld devices following secure password authentication is data encryption. We are referring specifically to the encryption of data that is kept on the device, and not the encryption of data that is in transit to or from the device via wireless or wired operations. The majority of wireless devices, or add-on peripherals that provide wireless features, are utilized to encrypt data to secure transmissions from one device to another. PDA Defense focuses on encrypting data that is stored on the handheld device itself, not encrypting data during the transmission from the device to any other destination.

Early on in the evolution of handheld device OSs, it became clear that the basic authentication provided could be bypassed, allowing access to the data on the device. For example, the Palm OS in pre-4x versions, as discussed previously, could not prevent debugging tools from compromising its authentication and downloading the contents of a device. According to Chris

Wysopal, technical director of research and development at @stake, a back door in the Palm OS allows anyone with developer tools to access information with the tap of a stylus.[1]

This issue was widely publicized in March 2001. As a result, a whole set of third-party applications, including PDA Defense, was developed and marketed to provide some level of protection against this identified threat. Because the system-locking mechanism could be bypassed, the encryption of sensitive data was the best defense against this threat.

Although that particular security hole is not a concern in the new handheld OSs today, other methods can bypass strong authentication in an OS, including gaining chip-level access to memory. For this reason, encryption must complement authentication as part of an overall handheld security plan.

How PDA Defense Handles Encryption   PDA Defense provides encryption protection that is transparent to the end-user. Any data selected for encryption is automatically encrypted when the device is locked, and it is decrypted when either the device is unlocked or the relevant data is accessed via the calling application (encryption "on demand" or "on-the-fly"). Thus, the end-user should never view or be able to see any encrypted data. However, if someone should bypass the authentication measures to obtain access to raw data on a locked device, the hacker would only find encrypted data.

A critical element in data encryption is how the encryption key is handled. If the encryption key is stored or accessible in any way, the encrypted data will be weak and vulnerable. PDA Defense employs a method that keeps the encryption key well protected from potential or would-be hackers. Essentially, when the device is locked, the encryption key is not stored on the device in any manner. When the user correctly enters his or her password, PDA Defense automatically generates the encryption key and keeps it available in a protected format while the device

---

[1]White, Aoife. "Palm PDA threat to network security." *Network News*, 15 March, 2001.

remains unlocked; thus, any data needed can be transparently decrypted.

When the device is locked by PDA Defense, the encryption key is used to automatically encrypt any data that was decrypted while the device was unlocked. PDA Defense keeps track of all the data selected for encryption and knows what needs to be encrypted once the device is locked. After encryption and prior to shutdown, the keys that were generated when the device was unlocked and used for encryption are destroyed.

Static Versus On-Demand Encryption Depending on the handheld device OS and the physical data storage medium, PDA Defense will do one of two things: It will either decrypt all relevant data at once when the device is unlocked, or decrypt it on demand after the device has been unlocked and an attempt is made to access the data by an associated application. This is an important feature for improving the usability of the device. In early versions of PDA security applications, most applications simply decrypted all the data when the device was unlocked. With the memory capabilities increased significantly today, and the variety of data-storage media available, being forced to decrypt all protected data immediately when the device is unlocked will not provide an acceptable performance for the vast majority of handheld device users. It will also detract significantly from the handheld device's ease of use.

Early on in its evolution, PDA Defense developed a method for the decryption of encrypted data only upon being accessed. Additionally, PDA Defense early on enabled specific databases or files to be selected for encryption so that performance could be further improved by not encrypting data that was not intended for protection, such as data that is used frequently.

For instance, healthcare patients' primary concern is ensuring that applications and associated databases that contain sensitive medical information are encrypted. However, physicians don't necessarily want to see the performance of their handheld device deteriorate just because they need to access some personal applications and data. Thus, PDA Defense enables the encryption of only the data that needs to be encrypted, and in many cases, that data is only decrypted when it has been accessed. Figure 6-14 shows PDA Defense selectively encrypting not only a specific application's data, but specific data files. For exampmle, in Figure 6-14, the specific data file "doc3" is selected for encryption under "Documents while the file titled "Untitled" is not selected.

PDA Defense's Data Encryption with Palm OS   Palm OS manages data in relationship to associated applications, and PDA Defense enables administrators to selectively encrypt application data. This includes any third-party or custom applications the organization has deployed. PDA Defense can also encrypt data stored on external storage cards that are compatible with a *virtual file system* (VFS). In order to protect data on a card with Palm OS devices, the end-user must initialize the card for protection. This initialization procedure requires that the user specify a password for the card. This password is unrelated to his or her user password for the device. However, the password is directly related to data stored on the card and protected by PDA Defense.

FIGURE 6-14.   Encryption selection table for Palm OS showing component files that can be selected for encryption.

After initializing the card by specifying a password, as illustrated in Figures 6-15 and 6-16, the user essentially specifies a "volume" of size that data will remain encrypted on at all times. It is possible to keep a protected volume on a storage card that

FIGURE 6-15.   Card manager defining a password.

FIGURE 6-16.   Card manager dialog box.

also contains unprotected data. For instance, after a user has defined the PDA Defense volume on a card that also contains other data that is not included in the volume, only the data in the protected volume requires the card password. PDA Defense encrypts or decrypts data on-the-fly as the card is read from and written to. If someone else takes the protected card and tries to read it in his or her Palm OS device, this person will not see any of the protected data. If he or she tries to view the data in a card reader on a desktop, the data will be encrypted.

After the card has been protected, each time the user inserts the card into his or her device, PDA Defense will automatically prompt them for the card password, as illustrated in Figure 6-17. The memory card feature enables some flexible alternatives for organizations wanting to share cards among many handheld device users. If someone else inserts a protected card, as long as he or she has the card password, the user can view or access the protected data.

PDA Defense's Data Encryption with PocketPC Depending on the data storage method used by PocketPC handheld devices, PDA Defense uses both decryption on demand and

FIGURE 6-17.   Memory card password prompt.

decryption upon device unlock. The choices for data encryption protection include:

- Databases
- Volumes
- Folders

Databases are formatted in a specified way so that certain PocketPC applications can access and store complex data formats. The PIM applications would be an example of PocketPC applications that use databases. In most cases, all the databases selected for encryption must be decrypted when the user unlocks the device. Therefore, some PocketPC applications that store large amounts of data in their database could affect how long it takes to unlock the device if the database has been encrypted. The PIM application databases are the exception here. With the release of PDA Defense v3.1, databases for standard PIM applications on PocketPC devices are only decrypted when the user attempts to access them.

Folders are essentially the same as what users experience on their desktop machines. Individual folders at the bottom of the folder tree, or parent folders (including all subfolders), can be selected for encryption protection. When a folder is selected, all the files in the folder are selected for encryption protection, and all the data selected is decrypted each time the device is unlocked. Thus, choices must be made carefully when selecting folders to encrypt because the wait time when the device unlocks could be significant if 30MB of data are selected for encryption. Also, care must be taken to not encrypt programs or any other system resources that might conflict with encryption. Figures 6-18 and 6-19 show what the end-user sees when selecting encrypted databases and folders on a PocketPC-based handheld device.

The third category of data storage on PocketPC devices is a volume, which is essentially a virtual folder defined by the end-user. Volume encryption on PocketPC handheld devices is handled the same way whether the volume is defined as a virtual

FIGURE 6-18.   Databases selected for encryption.

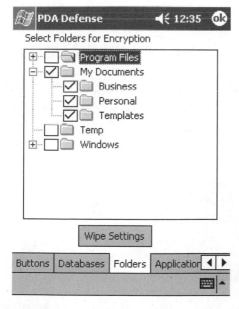

FIGURE 6-19.   Folders selected for encryption.

folder in *random access memory* (RAM) or on a storage card. In
both cases, a volume defined for encryption protection permanently keeps the data encrypted. The files are encrypted and
decrypted on-the-fly as data is accessed in the volume.

Unlike all other encryption settings for Palm OS and PocketPC devices, volumes on PocketPC devices cannot be defined
by the administrator; they are defined by the end-user, who has
complete control over them. Figure 6-20 demonstrates how the
end-user can define a volume.

One Enterprise used encrypted volumes on external storage
cards to give their users access to confidential information. The
portability of the storage cards allows the organization's staff to
share the card and the information. However, the information
can only be viewed in handheld devices that have PDA Defense
installed and only after the end-user enters the correct password
to authenticate the protected volume on the storage card, as
illustrated in Figure 6-21.

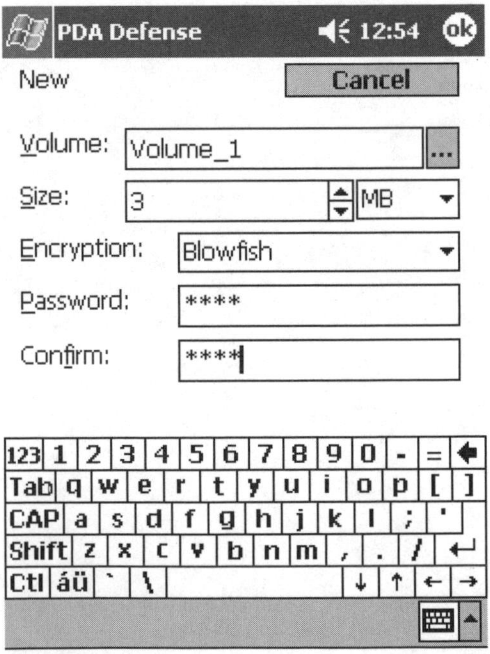

FIGURE 6-20.   Volume definition dialog box.

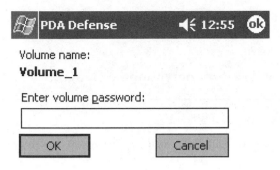

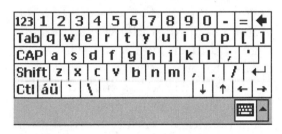

FIGURE 6-21.   Password request when a protected card is inserted into a PocketPC device.

PDA Defense's Data Encryption with BlackBerry RIM   PDA Defense allows standard or default PIM databases to be encrypted on BlackBerry RIM devices. These include mail, calendar, contacts, and memo databases. At the time of this writing, PDA Defense supports RIM devices running the RIM OS v2.0/2.1.

PDA Defense encrypts the data stored on the device when the system is locked. When the user successfully enters the password to unlock the device, PDA Defense automatically decrypts all databases that have been encrypted. For the RIM, no decryption on demand is available.

Encryption Algorithms Available   PDA Defense provides a selection of encryption algorithms that administrators can choose from for their end-users' devices. The encryption algorithms include:

- Blowfish with a 128- or 512-bit encryption key length.
- *Advanced Encryption Standard* (AES) with varying encryption key lengths from 128 to 256.
- A custom, fast algorithm utilizing an encryption key 128 bits in length.

The Blowfish and AES algorithms are well known and used by many security products. The AES algorithm, defined by the U.S. federal government's *Federal Information Processing Standard* (FIPS), utilizes the Rijndael algorithm. PDA Defense's fast algorithm is simpler in design and intended for use with information that organizations want to protect with some form of encryption that meets their needs. Designed as such, it may be easier to crack if given to someone with the proper training, tools, and time. PDA Defense provides a variety of encryption algorithm choices and varying key lengths so that each organization can design a custom policy file that works best for them.

A number of factors make up the encryption or decryption performance. These factors include:

- Algorithms.
- Encryption key length.
- Format of data storage or method.
- Media used.

The organization must decide which information will be encrypted as well as how it is to be stored before choosing the algorithm. More security-conscious organizations, such as government groups, may strictly use strong algorithms with the longest encryption key length available. However, many organizations will want to carefully consider the size of the databases they require their users to keep encrypted. For instance, large databases on external storage cards or on Flash *read-only memory* (ROM) will encrypt or decrypt much more slowly than similar sized databases in RAM or internal device memory. Because the length of time it takes to encrypt or decrypt can vary so widely, it is best that the administrator or IT group experiment with different encryption settings to get an idea of what their users' experience will be. They will want to avoid implementing policies that have an adverse impact on their end-users' handheld usability experience.

Encryption Performance as a Factor in Decision-Making Encryption performance varies based on three key variables: the *central processing unit's* (CPU) processing power, the algorithm and key strength, and the size of storage being encrypted and decrypted. Because encryption algorithms basically employ mathematical formulas to execute their tasks, they are by nature very computational and CPU bound. Additionally, because the complexity of the mathematical computations that must be performed on the data depends on the key strength and the algorithm, the key bit length and the algorithm form the second major factor. Finally, because each byte must be processed as a part of the computations, obviously the number of bytes being encrypted or decrypted provides the final component to determining the performance. Once the three components are selected, the particular software solution can do little to minimize the time it takes to execute the task.

A performance benchmark for the PDA Defense algorithms should be set up to evaluate how various encryption scenarios will affect the user. Because of the wide variety of hardware implementations available and the increasingly large number of memory configurations, this benchmarking should be established before rolling out a policy to your user community.

The following lists provide a briefly overview of some platforms' general performance. Using a 6,000-record address database (2.7MB), these performance measurements were reproduced in multiple test runs:

- Palm OS 5 (Tungsten T):
  - Asynchrony fast encryption algorithm: 7 seconds
  - Blowfish algorithm: 13 seconds
- Palm OS 5 (Tungsten T), not leveraging ARM processor:
- Asynchrony fast encryption algorithm: 7 seconds
- Blowfish algorithm: 26 to 30 seconds
- Palm OS 5 (Tungsten T), AES version included:
  - Asynchrony fast encryption algorithm: 7 seconds
  - AES algorithm: 90 seconds
- Palm OS 5 (Tungsten T), RSRC version included:
  - Asynchrony fast encryption algorithm: 7 seconds
  - RC4 algorithm: 13 seconds
- Palm OS 4 (Palm M125)
  - Asynchrony fast encryption algorithm: 13 seconds
  - Blowfish algorithm: 55 seconds

Determining Encryption Policies   The encryption options within PDA Defense provide numerous choices so that policy files can be customized to a company's various security requirements. Once the organization has chosen which type of encryption policies to set for each group of users, the administrator can utilize the encryption settings in the PDA Defense Policy Editor as part of the policy file-customization process. The administrator controls all encryption settings within PDA Defense, and the end-user cannot view or change any of the encryption settings specified by the administrator.

Encryption Selections for Palm OS Handheld Devices   For Palm OS handheld devices, the administrator has a number of options for identifying the data to be encrypted. Figure 6-22 shows the encryption options in the PDA Defense Policy Editor, and Figure 6-23 displays how the administrator can select which databases are specified for encryption protection.

All encryption settings for Palm OS devices reside on the Palm OS tab in the Policy Editor. Figure 6-22 shows that the administrator has selected all records to be encrypted. The

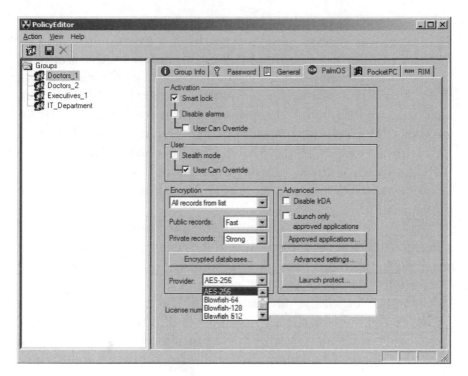

FIGURE 6-22. Palm OS tab with encryption settings.

FIGURE 6-23. Protected databases dialog box with applications checked for encryption protection.

other encryption options selected include the fast algorithm for public records and the strong algorithm for private records. Private and public records are those designated by the standard Palm OS applications (AddressBook, Calendar, MemoPad, and ToDo List). The public/private feature is strictly a Palm OS feature that was their initial implementation for securing certain records. PDA Defense has been designed to be compatible with this feature. Finally, the administrator has chosen the Blowfish-128 algorithm as the strong option, although you can see in the drop-down list some of the other algorithms available to the administrator.

In Figure 6-23, the administrator has chosen to encrypt all the records selected from the database list. The databases selected in the figure include the Address book database and the MemoPad database. Their selection is denoted by the check-boxes to the left of the database name. If the administrator had followed the steps earlier to create a set of template policy files, he or she would also be able to see other databases associated with third-party or custom applications residing on their users' handheld devices.

Encryption Selections in PocketPC Handheld Devices

For PocketPC handheld devices, the administrator has the ability to select some of the encryption options for certain database types. As mentioned previously, databases, folders, and volumes on PocketPC handheld devices are available for encryption.

The administrator has the ability to select which databases and folders will be selected for encryption as part of the policy file. As a result, the end-user cannot change these choices, similar to how the end-user cannot change any of the encryption database selections on Palm OS handheld devices. However, the administrator does not specify the volumes chosen for encryption. As described previously, the end-user defines the volumes where he or she would like to store encrypted data.

In the PDA Defense Policy Editor PPC tab, as illustrated in Figure 6-24, the administrator can click the Encrypted databases and Encrypted folders buttons to select which databases and folders are chosen for encryption protection.

As discussed earlier in this chapter, in the "Creating the Template Policy File: Copying a Set of Baseline Applications and Database" section, the administrator must import an image of the third-party or custom applications to get their associated databases as well as any custom or specific folders. Figures 6-25 and 6-26 illustrate the Protected database and Protected folders dialog boxes.

If the organization wants to select additional applications or custom folders for encryption, the administrator must use the Import feature so that the Policy Editor can get an image in the

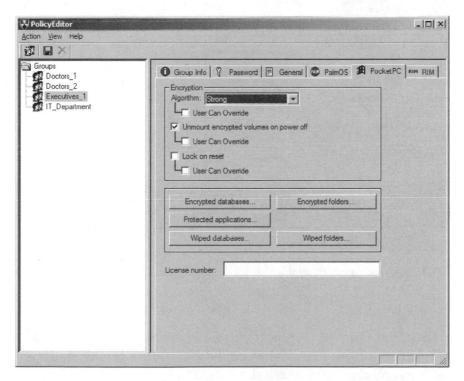

FIGURE 6-24.   Pocket PC tab on Policy Editor.

FIGURE 6-25. Protected databases dialog with third-party application database.

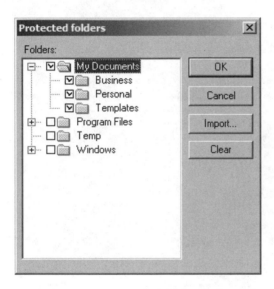

FIGURE 6-26. Protected folders dialog with additional folders defined on the PPC device.

policy file. Once this import process is complete, the administrator can simply check the databases and folders that should be encrypted on end-user's devices.

The other encryption option that the administrator can preset is the choice of a strong or fast algorithm when the selected databases, folders, and volumes are encrypted. For PocketPC devices, only the Blowfish encryption is available with PDA Defense Enterprise version 3.1. Additional algorithms will be made available in future releases.

Encryption Selections in BlackBerry RIM Handheld Devices For BlackBerry RIM handheld devices, the administrator can select which of the primary PIM databases receive encryption protection when the device is locked. Figure 6-27 shows the encryption settings available to the administrator on the PDA Defense Policy Editor.

In Figure 6-27, the administrator has selected to keep the Tasks and Address book databases encrypted on his or her end-user devices. Similar to PocketPC devices, the administrator can choose whether to use the fast or strong algorithms for data encryption. In the 3.1 release, Blowfish is the algorithm currently available for the strong choice.

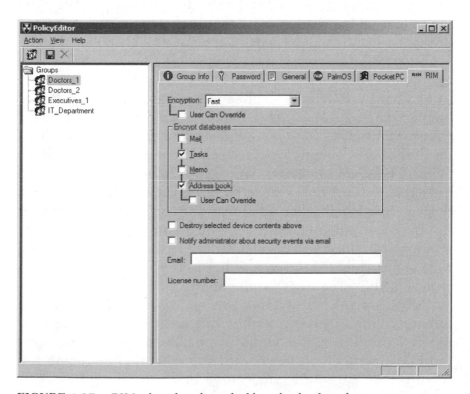

FIGURE 6-27. RIM tab with tasks and address book selected.

Customizing Policy Files: Determining Fail-Safe Policies (Bit Wiping)

The third major level of protection for handheld devices following secure password authentication and data encryption is the use of fail-safe bit wiping. The bit wiping in PDA Defense is also known as the "bomb." Although bit wiping seems intimidating and can mean a total loss of data on the device, most users sync or back up their devices on a regular basis, mitigating the scope of data loss. Given the relatively easy ability to restore data from backups by a simple syncing process, fail-safe bit wiping becomes a reasonable security precaution. However, each organization must assess for themselves whether they should utilize or enforce bit wiping in their policy files. It is the responsibility of the administrator to understand how employees use their handheld devices, including the volatility of data on the devices in relation to the frequency of the syncing or backup of the data. Based on that knowledge, the administrator and Enterprise can decide how to handle bit-wiping options in their policy files.

The two primary methods for invoking the fail-safe bit-wiping options include the number of consecutive incorrect password attempts (attempts bomb) and the amount of time elapsed among syncing events (time bomb). Although the incorrect password attempts limit is available on all platforms, the limit for time elapsed among syncing events is only available on Palm OS and PocketPC devices.

Similar to other features, such as the autolock and time delay before locking, the administrator can specify whether or not the bit-wiping options are enabled and whether or not the end-user has access to enable or disable those options. Figure 6-28 shows the PDA Defense Policy Editor's General tab with a typical profile of settings for the bomb.

In Figure 6-28, the administrator has selected to set the attempts bomb to five attempts but not to activate the time bomb. In this case, the administrator has made the attempts-limit fail-safe method mandatory, whereas the time bomb is a feature that end-users can enable and set themselves. This might be an example of a profile that an organization wants to

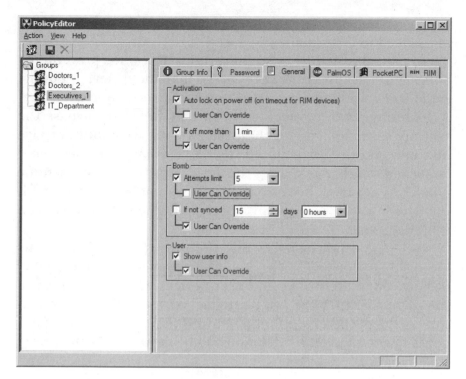

FIGURE 6-28. Bomb-related settings.

deploy to a group of users who do not sync regularly. Those users that do sync regularly can elect to enable the time bomb feature.

Password Attempts Limit (Attempts Bomb)   The password attempts limit, also known as the attempts bomb, is one of the original features of PDA Defense. PDA Defense was the first handheld security application to offer this feature in the early stages of the endpoint device security market. Although other security solutions require password authentication, many do not limit the number of attempts. This leaves those applications open to brute-force attacks.

Time Between Syncs (Time Bomb)   Because handheld devices are easily lost or stolen, a logical enhancement in one of the early releases of PDA Defense was the addition of the time bomb. When enabled, this feature assures that any selected data will be deleted after a specific period of time when the device is not synced to a desktop.

In fact, this feature is so powerful that it once prevented a police investigator from accessing a handheld device confiscated from a felony suspect. When the investigator contacted Asynchrony looking for clues as to how he might bypass the security to gain access to the device's data, we could not give him one because no planned backdoor existed. In fact, what happened was that as soon as the device was turned on, the time bomb was engaged, because the end-user enabled it to go off if the device was not synced in a period of time, which had been exceeded.

Although PDA Defense was unfortunately not put to its intended use for good here, hopefully you can see the parallel of how your Enterprise's end-user devices will be protected in the event that they are lost or stolen. With the time bomb feature enabled, the organization can rest assured that any confidential data selected for deletion will be destroyed in the preset timeframe.

Wipe Password   Another fail-safe measure incorporated into PDA Defense is the use of a wipe password. It is implemented only for Palm OS-compatible devices in the 3.1 release of PDA Defense Enterprise. Figure 6-29 shows how the user can set this option. The Advanced Options dialog box can be accessed from the main PDA Defense screen. The wipe password is enabled as soon as the end-user creates an alternative password in the appropriate field.

The wipe password engages the bomb as soon as the user enters the appropriate password in the lockout screen. It is an immediate wipe that takes place without any warning.

The entry of the wipe password can also be accomplished by the administrator in the PDA Defense Policy Editor. However, because of the nature of this feature, the password can be overridden or changed by the end-user. Also, the end-user will need to know what the administrator sets for the wipe password, because he or she should never use the same phrase for their PDA Defense password.

FIGURE 6-29. Palm OS's Advanced Options dialog box.

This feature was originally intended to satisfy a requirement in secure government facilities that needed to appropriately "clean" a device before allowing it to leave the secure facility.

Choosing the Databases Wiped by the Bomb The PDA Defense administrator must select which databases are wiped if the bomb is engaged. Due to the nature of permanent deletion from the device, by default nothing is selected for deletion until the administrator specifies it. Figures 6-30, 6-31, and 6-32 illustrate the dialog boxes where the administrator can select what is targeted for deletion by the bomb.

For the PocketPC devices illustrated in Figures 6-31 and 6-32, the wiped databases' and folders' settings can be accessed from the PocketPC tab in the Policy Editor. Similar to how an administrator must import any third-party or custom applications for encryption, the administrator must also import a template of what is on the device so that it appears in the selection screen for the bomb. Then, the administrator simply has to check the appropriate boxes next to the databases or folders he or she wants to wipe.

For the Palm OS policy settings in Figure 6-30, the screen where databases are selected for encryption is also used for

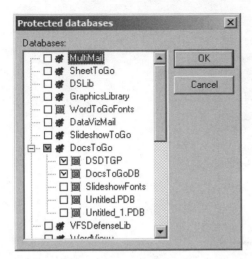

FIGURE 6-30. Protected Databases dialog box from the Palm OS tab in Policy Editor.

FIGURE 6-31. Wiped databases from the PocketPC tab.

selecting which databases are part of the bit-wiping process. To access the Protected databases selection list, the administrator must click the Encrypted databases button on the Palm OS tab of the PDA Defense Policy Editor. When choosing databases, the checkboxes in the list are those available for encryption,

FIGURE 6-32.   Wiped folders screen from the PocketPC tab.

whereas an X over a table symbol signifies that a database is selected for deletion as part of the bomb process. By clicking the table symbol next to the checkbox, the administrator can turn the red X on and off. Any settings in these screens will be saved as part of the PDA Defense policy files.

Remote Bit-Wiping on RIM Devices   One additional fail-safe feature in PDA Defense Enterprise version 3.1 works only for BlackBerry RIM devices. The administrator can create a policy file that will automatically engage the bomb process on a RIM device when the policy file is received via email. It uses the same policy file delivery process as all other policy settings. This feature will become standard as wireless devices proliferate.

In Figure 6-33, the administrator has not only selected the Tasks and Address book databases for encryption protection, but also the "Destroy selected device contents above" option. When the resulting policy file is received on the intended RIM device(s), the Tasks and Address book databases will be automatically deleted or wiped by the bomb process.

This powerful feature provides up-to-the-minute control over the destruction of remote devices. Similar to how the time

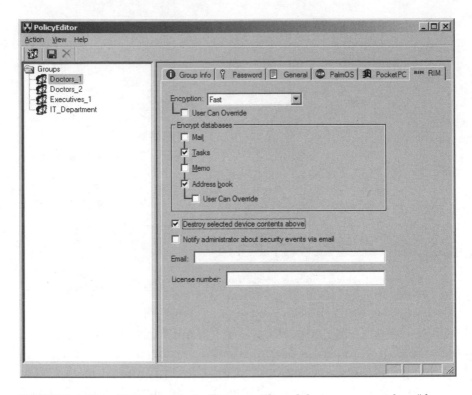

FIGURE 6-33. RIM tab with the "Destroy selected device contents above" box checked.

bomb ensures that data on a lost device is kept safe by deleting it, the remote bomb safeguards information on devices, enabling administrators to actively manage the bomb process.

Customizing Policy Files: Other Considerations for Policies

In addition to the three primary levels of protection, PDA Defense also offers a number of additional features to provide added security controls. Organizations can choose to implement some of these features, depending on their own needs and requirements.

APPROVED APPLICATIONS   In addition to providing secure authentication, data encryption, and fail-safe bit-wiping measures, PDA Defense also provides the capability to restrict which applications can be run on Palm OS-based handheld devices. An organization may want to allow only vertical applications or those that are strictly related to an employee's work to be run on handheld devices. For example, if a healthcare organization conducting clinical trials of a new drug deploys handheld devices, they may want those devices to be only used for applications directly related to the clinical trials. Also, to ensure the security of the medical information stored on the devices, restricting the introduction of new applications reduces the user's ability to introduce hacks and other programs that might intentionally or accidentally interfere with the security measures provided by PDA Defense.

Before selecting which applications to run on a device, the administrator must first create the template policy files by copying an image of which applications and databases are installed on the device. This step was discussed previously in the "Creating the Template Policy File: Copying a Set of Baseline Applications and Database" section. Once the administrator has imported the template policy files into the PDA Defense Policy Editor, he or she can then check the "Launch only approved applications" box in the Advanced section of the Palm OS tab, as illustrated in Figure 6-34.

Checking this option and selecting the applications the organization wants to restrict on each handheld device in the appropriate group will apply this rule when the policy files are deployed. After the policy files are applied, any attempt by the end-user to launch an application that is not approved will be intercepted by PDA Defense and halted. Additionally, the end-users will not be able to install and run other applications that are not marked as approved to run.

Many organizations will have to carefully consider using this form of restrictive policy that includes limiting which applications can be run on devices. If many of the devices are owned by the corporation and are intended to be primarily devoted to business tasks, this is a powerful policy. It can also help users avoid

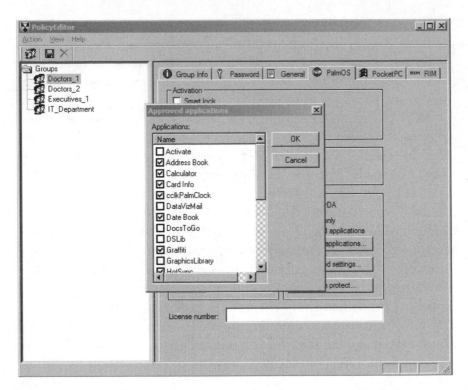

FIGURE 6-34.   Palm OS tab with approved application option checked.

the temptation of downloading unrelated applications that might cut into an employee's productivity, a temptation many Palm OS users have. However, if a significant number of employees receiving a policy like this actually use their personal devices, the organization might have to consider other alternatives.

Application Launch Protection   Another additional protection feature includes application launch protection. This requires the user to enter a password when he or she attempts to launch an application. This authentication takes place in addition to unlocking the device itself. Some organizations feel more secure if they require additional levels of authentication to launch certain applications. The launch protection feature is available on Palm OS and PocketPC devices.

Similar to the "Approved Applications" scenario, the administrator must first create the template policy files by copying an image of the applications and databases installed on the device that he or she wants to apply the policies to. For PocketPC devices, the administrator will need to follow a process similar to importing databases and folders. Once the administrator has imported the template policy files into the PDA Defense Policy Editor, he or she can click the "Launch Protect" button, as in Figure 6-35. He or she can also select the "Protected Applications" button from the PocketPC tab, as illustrated in Figure 6-36.

Application launch protection is commonly used by organizations in the healthcare field. Any sensitive or private medical record can use the additional level of security or protection. In the event a device is found before an automatic lock has been activated, the launch protection would require the password before any information could be accessed by an unauthorized user.

Disabling Infrared Data Association (IrDA)   With respect to wireless technologies, a convenient and common feature of handheld devices is the use of the IR port to send or receive "beamed" data short distances. Like other handheld features,

FIGURE 6-35.   Palm OS tab with Launch Protect dialog box open.

FIGURE 6-36.   PocketPC tab with the Protected Applications dialog box open.

this convenience comes with the threat of unauthorized access to confidential information on the device. A number of security-conscious organizations are so concerned about infrared transmissions of information that they have banned the use of handheld devices.

One of the requested features in handheld security solutions is the use of IR ports that can be disabled. When a device is locked and unlocked, the IR port represents a means to send and receive proprietary information if it is not secured. Additionally, the IR port can be utilized to deliver Trojan horse and other attack applications to a device, potentially bypassing security countermeasures if not disabled.

PDA Defense Enterprise not only locks the IR port when the device is locked, but it also offers the administrator the option to disable the IR port for some devices when the device is unlocked. Figure 6-37 displays the PDA Defense Policy Editor allowing the administrator to disable the IR port. The administrator selects this option to disable IR along with the rest of their policy settings. He or she then saves and deploys the policy files to their end-users.

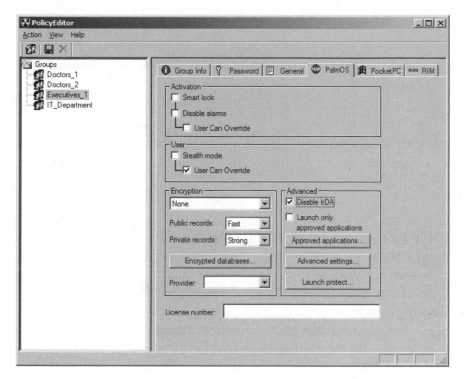

FIGURE 6-37. Palm OS tab with the Disable IR port option highlighted.

Additional Features Planned for the Future A number of additional features are slated for future releases to provide organizations with additional flexibility in their PDA security solutions. With the field of handheld devices changing rapidly, new technologies are continuously being introduced every several months, causing new threats to information security. Some of the more recent technologies involve primarily wireless methods, such as:

- Bluetooth-enabled devices.
- 802.11-compatible devices and add-ons.
- *Global positioning system/general packet radio service* (GPS/GPRS) cellular network access.

- *Code Division Multiple Access* (CDMA) cellular network access.
- Increasing use of pager/radio networks for low-bandwidth applications.

As these technologies gain more widespread acceptance and become ubiquitous in all new handheld devices, security solutions will evolve to meet the challenges of using these new technologies.

DEPLOYING PDA DEFENSE AND POLICY FILES

One of the vital components of implementing a security solution for handheld devices is the capability to deploy it across an organization. In some cases, this deployment must be done over a widespread geographic area. Some of our customers implementing PDA Defense have utilized it across hundreds and thousands of miles.

PDA Defense was designed and built to be compatible with not only the native syncing software that ships with the various handheld devices and their platforms, but also third-party syncing solutions. A number of these solutions were described earlier in Chapter 3, *"The Power Resource Guide to Understanding Where Security Must Be Achieved."* They provide additional capabilities when deploying software and handheld devices over a geographically dispersed network. Some of the most popular third-party syncing solutions include XTNDConnect Extended Systems and Afaria by Xcellent. These solutions utilize "channels" to deliver software to individual users, and they add other valuable services such as monitoring user sync sessions and providing information for the entire Enterprise or a specific group of users. The PDA Defense software and the organization's customized policy files can also be easily deployed through these distribution software solutions.

Once the administrator has completed customizing the policy file settings, he or she can save the policy files and deploy them to users. PDA Defense is designed so that the administrator can deploy just a policy file update to PDAs already in the field or provide both a policy file update and the PDA Defense client programs as well.

Figure 6-38 demonstrates how the administrator can easily save the policy files for the appropriate user group by simply selecting the platform(s) that applies to those users. The policy files are saved on the administrator's desktop where the Policy Editor executes. They are saved as "pdb" data format files for Palm OS devices, "pds" format for PocketPC devices, and "pdr" format for RIM devices. As part of the policy file generation, the administrator is prompted to save the policy information gener-

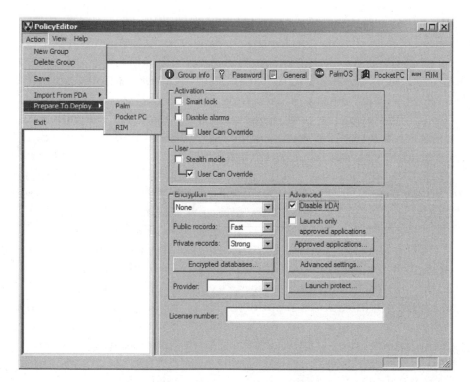

FIGURE 6-38.   Policy Editor with the "Prepare to Deploy" menu item selected.

ated for the group selected. The administrator is also prompted to save if he or she switches from one group to another or upon exiting the application. Once the administrator has generated all the appropriate policy files using the PDA Defense Policy Editor, he or she is ready to deploy those policies with the PDA Defense software.

Updates to policy files are generated in the same manner as the initial policy files. The administrator can launch the Policy Editor and open the appropriate group whose policy settings need to be modified. Once he or she is satisfied with the modifications, the administrator simply saves the files, as illustrated in Figure 6-38, and distributes the policy files via the network or syncing solution.

Because security requirements change over time, the ability to easily and efficiently update policies is an important feature. Additionally, if the organization rolls out a custom application to a group of users, the administrator may need to update the encryption, fail-safe, or another security measure related to that new application. As mentioned earlier, the administrator password can easily be changed as part of a periodic policy file update so that the administrator can keep the settings secure.

When designing and implementing a security solution that imposes additional controls and restrictions on end-users, some level of resistance will always take place. However, with the many flexible options, each organization can design a set of policies that best fits their organization. They are free to set tight controls for one set of users and give other users who don't require the same level of security a wider range of latitude.

PDA Defense provides a flexible solution to use across an organization. It is inevitable that some users will resist the imposition of any limitations or security on their devices. However, it is the organization's responsibility to protect and secure any proprietary information regardless of how individuals might feel. In the end, if the administrators choose their settings appropriately, most end-users will not feel the security significantly detracts from their handheld user experience.

FROM EVALUATING RISK TO ENFORCING POLICIES: WHAT'S NEXT?

We covered a lot of ground in Section 2, *"Handhelds in the Enterprise: When, What, and How,"* from assessing risk to creating and enforcing policies on devices in the field. This high-level overview of managing the increased risk introduced by PDAs hopefully fit within a context you are already familiar with in the area of security or IT management. However, this chapter on understanding how to manage risk in the coming age of ubiquitous computing may have left some still hungry for a deeper look at achieving PDA security objectives.

The next section takes a more technical look at the key components behind executing a technical solution to mitigate PDA risk. From an examination of the key issues around the various OSs to some of the details around authentication and encryption, we will attempt to demystify and focus your efforts around the key technologies you should better understand.

THE TECHNOLOGY OF PDA SECURITY

CRYPTOGRAPHY, PASSWORDS, HACKING, AND MORE

Chapter Seven

UNDERSTANDING HANDHELD SECURITY IS LARGELY ABOUT THE OPERATING SYSTEM

In this chapter, we will start with an in-depth conversation about the key technical elements of achieving security on a *personal digital assistant* (PDA). In Section 2, *"Handhelds in the Enterprise: When, What, and How,"* we reviewed risk management and policy, and discussed how to enforce standards and policies through the use of PDA Defense as an example of policy enforcement software tools. As we move into Section 3, we begin a deeper and more technical examination of PDA security, focusing on the technical elements of passwords and encryption, along with a discussion of the major operating systems in use. We start with a technical review of password authentication and encryption.

The premise behind addressing the operating systems is that while each PDA manufacturer has its own vision of implementing security, the operating system dramatically sets devices apart from a security perspective. A wide range of operating systems are in use on devices including, but not limited to, Windows CE (PocketPC, PocketPC 2002, WinCE .NET), Palm OS (v4.x, v5.0), Symbian OS, *Research in Motion* (RIM) OS, Linux, and Java VM. But the PDA market is largely driven by devices with one of the following two operating systems: PalmOS or Windows CE. These devices encompass almost 90 percent of the PDA market.

It is important to develop an understanding of the default security implemented on Palm and Microsoft devices. While security implementations can vary within each operating system by device, in general a locked device will require a password or *personal idenftication number* (PIN) for entry. While newer devices and third-party products have implemented thumb fingerprint, smart card, and other methods of entry, the majority of locked device are still entered with a PIN or password. Devices with the Palm or PocketPC 2002 operating system prompt the user to enter a password to unlock the device. New PocketPC 2002 devices such as HP iPAQ 5400 series have fingerprint readers for enhanced security. Some PDAs, such as the HP Jornada 720, have smart card reader slots. Built-in security implementations differ slightly between OS releases (WinCE, PalmOS, Symbian OS, and so on), and even, at times, between OS sub-versions.

So we will begin this technical conversation by discussing the major components of security on the PDA, and conclude with a review of the major operating systems for PDAs in the market.

PDA SECURITY AREAS

Before discussing specific operating system issues, we will discuss key elements of PDA security, including the following six areas:

- Device access authentication.
- Network connection security.
- Data storage security (data encryption).
- Resistance to intruder penetration.

- Cryptography.
- Access to device storage bypassing the OS.

Device Access Authentication

Device access authentication allows only authorized users access to the PDA. The most common authentication device is the password, which will be discussed in detail in this section. The other less common authentication methods, such as smart cards, will be touched upon at the end of the section.

All passwords consist of a string of characters at least one character long. An alphanumeric password is a string that consists of alphabetic, numeric and special characters. The PIN is a specific password that contains only numeric characters, often four characters long. The following points apply to all passwords, regardless of how they are composed.

The key objective of a password is that it cannot be cracked. Obviously, one-character passwords are not sufficient because another user can break the password by trying each character one-by-one. In the time a person can try hundreds of passwords, computers can create millions of attempts to try to break into the device. This is called a brute-force attack. Thus the use of a "strong" password is very important. An in-depth discussion of strong passwords follows.

Password Storage   To authenticate a user, the device must store password-related data to validate it against the user's entry. The simplest method is to store an original password on the device in standard memory, but keeping the password in clear text would allow for easy password recovery if the storage media were analyzed. Thus, the password should be stored using one of the following methods:

- Store scrambled password.
- Store password hash.

Store Scrambled Password   A scrambled password is one in which the various characters are stored in non-consecutive memory, often offset by a hard-coded value. The result is that a person looking for the password in memory will not find it by using a hex-editor or "strings" programs that pickout clear text hiding in a memory. Instead, the algorithm that scrambled the password is what must be reverse-engineered or otherwise understood to identify the password without requiring a brute-force attack. Unfortunately, a scrambled password results in weak protection since it relies upon the programmer, who likely is not a security expert and who likely will be implementing a poorly tested approach to create a form of encryption on the storage media. Cryptography relies upon thorough vetting of algorithms; developing custom cryptographic approaches without relying on well-tested public algorithms will generally lead to a weak solution. In short, hackers may reverse-engineer the program that retrieves the scrambled password, and simply recover the original password, allowing the device to be unlocked with ease.

Store Password Hash   Storing the password using the MD5 one-way hashing algorithm or some equivalent approach is more reliable than any custom password scrambling. A *hash* is created using some algorithm on the password itself, but the password is unlikely to be recovered from the hash. At this time, cryptographers use one-way mathematical functions to produce the hash from the original password string. A "one-way function" means that it is easy to calculate the hash from the original password, but difficult to calculate a password from the hash. At this time, MD5 and SHA-1 hash algorithms are most often used. The MD5 algorithm generates a 128-bit hash from any string, and SHA-1 generates a 160-bit hash from the initial string with arbitrary length.

The irreversible nature of the hash does not protect it from brute-force attacks, however. Commonly available tools allow hash attacks based on libraries of hashed values that can be compared hash to hash. For example, if I hash the English dictionary and then compare my hash dictionary against the hash value, I will find a match if the password was originally in the dictionary. Generally speaking, this—like all password authentication examples—will ultimately come back to the strength of the original password. However, to better evaluate the risk of a brute-force attack, we can calculate the reliability of a password hash against a brute-force attack. Table 7-1 allows us to review a couple of scenarios. Let us imagine that we have a computer that checks one million passwords per second.

These numbers are very big, so big, in fact, that the total number of seconds required to break the password exceeds the age of the universe (~20 billion years!). Even if we use a faster computer that is able to enumerate one billion passwords per second, the total time would be extremely long. Thus the password hash is very reliable, assuming we don't allow for any shortcuts such as a simple dictionary hash attack.

Password Attacks and Choosing the Right Password
Because PDAs are generally protected by passwords, you need to ensure that your user community chooses the right password. The practice of choosing a simple one-word password using a name, spouse's name, friend's names, pet's name, birth date, or phone numbers, for example is not advisable. A simple one-word password is easy to remember and recover without assistance. Many people know a user's name, spouse's name, pet's name, telephone numbers, and so forth, and can guess the password.

Brute-Force Attack Very simply, what a brute-force attack does is try all possible values for a password until it hits the one that produces the hash. Obviously, the more complex the pass-

TABLE 7-1. Strength of Password Hash

	Length (bits)	Length (bytes)	Number of possible values	Time to break password (years) 1 million passwords per second	Time to break password (years) 1 billion passwords per second
MD5	128	16	$2^{128} \sim 3.4 \times 10^{38}$	$1.0 * 10^{25}$	1.0×10^{22}
SHA-1	160	20	$2^{160} \sim 1.5 \times 10^{48}$	$1.0 * 10^{34}$	1.0×10^{31}

word, the tougher it is for a brute-force attack to find the password.

Thus, a password should be "good"—not so simple that a brute-force attack will easily crack it. Password cracking programs often use well-known word dictionaries, which list words often used as passwords. Thus, the crack time of a poor password would be decreased from billions of years to several seconds. For example, "king" is not a good password, but Hgt6&_%dcA#k2 is. For foreign users, the practice of typing on another language keyboard is not effective because hackers know this practice, and dictionaries for this input method also exist.

Password Crackers   Password attacks do not require great skill on the part of the attacker. There are many easily available password recovery tools, such as Lophtcrack or LC4 (see www.atstake.com/research/lc/index.html). Figures 7-1, 7-2, and 7-3 illustrate some common methods available from the Lophtcrack tool.

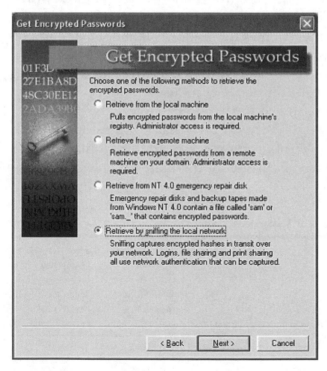

FIGURE 7-1.   Setting in which password hash or encrypted password should be retrieved.

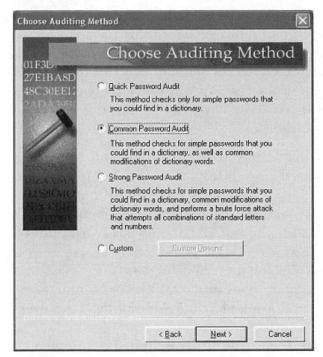

FIGURE 7-2.
Methods that should
be used to try to
recover password from
the hash.

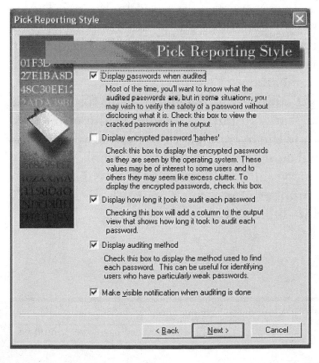

FIGURE 7-3.
Report settings.

These tools also allow recovery of passwords from the password hash using word dictionaries; algorithms for adapting to common dictionary word modifications; and brute-force attack using a limited/targeted character set. You can use these tools to check the strength of a password you want to create. If it is not strong enough, increase its length as well as the character set from which it is created.

A Mathematical Look at Password Strength and Structure

We can estimate how strong different password schemes are. Let's imagine that we use a computer that is able to check one million passwords per second. We will use the assumption that we should enumerate one half of all possible password values.

Simple-Word Passwords　Using a dictionary to perform a password attack provides for thousands of words. Let's assume that the typical dictionary contains 60,000 words. Our computer will enumerate all available words during the time illustrated in Table 7-2.

Concatenated-Word Passwords　This scheme provides a more reliable alternative. We choose several words and join them, using special characters, as illustrated in Table 7-3. As a result, even using simply generated passwords from an existing dictionary by joining two or more words with the inclusion of punctuation, we have created a relatively "strong" password. This is known as a concatenated word.

To ensure that we address more general brute-force attacks not driven by a dictionary, we must also address the number of possible characters and the overall length of the password. If the password uses small and large alphabetic symbols, numeric symbols, and special symbols, the number of possible characters is 100 and the number of possible passwords becomes very large, as illustrated in Table 7-4.

TABLE 7-2.

$T = (60000/2)/1000000 = 0.03$ seconds: Very fast!

Using the example from Table 7-3, we can look at the minimum length of a password required to achieve the strength illustrated in the mathematical example of Table 7-3. Table 7-5 builds on the examples in Tables 7-3 and 7-4 to back into the target minimum character length of a password based on the method of word aggregation with the range of characters illustrated in Table 7-4.

If we extend this example to practical extremes of usability, Table 7-6 shows a final example based on concatenating five words, probably the most complex example of this type of password.

Finally, to ensure adequate length of passwords, the example in Table 7-7 illustrates the associated time to crack passwords based on their varying lengths.

TABLE 7-3.

$N = D^M * K^{M-1}$

Number of possible words in passwords = M

Number of possible punctuations characters = K

Number of possible password = N

Dictionary size = D

For example M = 3, K = 10, D = 60000.

$N = 2.16 \times 10^{16}$

Required time to crack

$T = (N/2)/1000000 = 1.08 \times 10^{16}$ seconds > 340 years

TABLE 7-4.   Number of Passwords with Use of Larger Character Set

$N = P^L$

Number of possible passwords = N

Number of possible characters = P

Number of characters in password = L

TABLE 7-5. Target Minimum Length Exercise

$N = D^M \times K^{M-1}$

Number of possible words in passwords = M

Number of possible punctuations characters = K

Number of possible password = N

Dictionary size = D

For example M = 3, K = 10, D = 60000.

$N = 2.16 \times 10^{16}$

$L = \log(2.16 \times 10^{16}) / \log P$

Number of possible characters = P

Number of characters in password = L

For example P = 100

$L > \log(2.16 \times 10^{16}) * \log 100 = 8.16$

L should be > 8 characters

TABLE 7-6. High-Difficulty Example

M = 5 words

$N = 7.78 \times 10^{27}$

Required time to crack the password

$T = (N/2)/1000000 = 3.89 \times 10^{21}$ seconds $\sim 1.23 \times 10^{14}$ years.

For comparison estimated age of observed universe is about 20 billion (2×10^{10}) years.

TABLE 7-7. Understanding the Implications of Password Length

Using all possible printable ASCII characters including special characters = 100

$N = 100^L = 10^{2L}$

Where L is password length.

L	N	Crack time
8	10^{16}	160 years
15	10^{30}	1.6×10^{16} years
20	10^{40}	1.6×10^{26} years

Password Generators   One method of ensuring your user community operates with adequate passwords is to adopt a rules-based "password generator." A password generator is a program that generates random passwords using defined characteristics such as length, password mask, characters set, and so on. An advantage of using password generators is producing a password that cannot be cracked using word dictionaries. The disadvantage is that the resulting passwords are random unpronounceable strings, which are hard to remember and often lead to users writing them down in one or more places around home or the office. There are a lot of available password generator programs, even online, where you can get passwords by pressing a button on the Web.

Table 7-8 illustrates source code from a sample password generator. If executed, the code in Table 7-8 will generate passwords similar to the examples shown in Table 7-9.

In conclusion, because most devices utilize password authentication, it is important that the users choose a password that cannot be easily cracked. Third-party software products often help in forcing the user to choose a good password, as we have discussed in the policy enforcement example of Chapter 6, *"How-To Guide: Asynchrony's PDA Defense, Palm's Choice for the Enterprise."*

Other Authentication Methods   Other authentication types include smart cards and biometric methods. Smart cards provide reliable authentication by using data stored inside the card or by using reliable authentication based on the calculated secrets using *central processing units* (CPUs) inside smart cards. Biometric methods use unique human characteristics such as fingerprints, iris image, retina image, human voice, and face image. Both methods are more secure than password authentication, but they are also more expensive and potentially less convenient (e.g., having to swipe a Smart Card every time a user wants to use his or her PDA). Also, as we discussed in Chapter 5, the user must manage the issue of false positives/negatives.

Biometric Devices   Recognition systems usually work in two modes: registration and verification. During registration, certain biometric characteristics are measured to create a "template"

TABLE 7-8. Password Generator Sample Code

```
/* This is a sample program that uses a pseudo-random
generator from a standard C runtime library to produce
random passwords.
*/

#include <stdio.h>
#include <stdlib.h>
#include <string.h>
#include <time.h>

#define PASS_LEN      10

char charset[] = "ABCDEFGHIJKLMNOPQRSTUVWXYZ\
abcdefghijklmopqrstvwxyz\
01234567890\
~!@#$%^&*()_+|-=\\[]{};':"",.<>/?";

void GenPassword(char* passbuff, size_t pass_len)
{
size_t i, charsetlen;
charsetlen = strlen(charset);
for(i = 0; i < pass_len; ++i)
        passbuff[i] = *(charset + rand() % charsetlen);
passbuff[i] = '\0';
}

int main(int argc, char* argv[])
{
time_t curtime;
char random_password[PASS_LEN+1];
srand(time(&curtime));
GenPassword(random_password, PASS_LEN);
printf("Password is %s\n", random_password);
return 0;
}
```

TABLE 7-9. Password-Generated Examples

```
Password is SFN2E^H5]+
Password is y8Rp&,8U<s
Password is G'L<5C4_DV
```

which is then used by the system for verification of the user when the user's biometric characteristics are represented at a future point. This template is usually stored in an internal database. We will describe this process of Registration or Template creation and Verification or Template comparison in more detail at a later point in this chapter.

Registration   First of all, each user needs to be registered by a system administrator as an authorized user. A hardware device called a sensor, which is usually placed on the front panel of the biometric authentication device, measures biometric characteristics. A biometric sensor produces a signal that corresponds to the value of determined characteristics. For example, if a sensor is a microphone, it produces a signal with amplitude (electric current or voltage) dependent on the recording of the voice.

Most biometric devices produce analog signals and output signals that are converted into digital form. Thus, the next step in this chain should be an analogous-digital converter. An analogous-digital converter accepts analog signals and outputs a digital stream of the original analogous signals. Often, biometric systems use final digital images instead of raw analog biometric signals. A digital biometric image is compressed to produce a biometric template. Almost all biometric systems use several templates for each person to reflect the various natures of biometric identification variations.

Verification   After registration, a user's biometric characteristics are used for authentication. During authentication, a sensor records human biometric characteristics that are converted into the biometric digital image and compared with the user's biometric template. Biometric characteristics always contain variations; biometric systems cannot demand exact matches between a recorded template and the last recorded biometric characteristic. Biometric characteristics are treated as valid if they fall into a range of values. A comparison algorithm determines whether a biometric characteristic is valid.

The system administrator determines the validation range. If the range is too large, the biometric authentication will not be reliable. However, if the range is too small, there is a chance that the valid user will not receive authorization.

Templates   One of the biometric authentication key features is the template. User templates are placed into a template database. This database must be protected like passwords. The template size is different for each biometric system.

Some biometric systems use templates for verification, but some of them use templates for identification. A *biometric identification* system compares characteristics with all the templates in the database. *Biometric verification* compares current characteristics with only one template. Verification systems work faster than identification systems because they don't need to compare records with all the templates in the database.

Recognition Methods   Each person has a set of unique biometric characteristics. There are several recognition methods:

- Fingerprint recognition
- Optical recognition
- Handwriting sign recognition
- Face image recognition
- Voice recognition

FINGERPRINT RECOGNITION   Fingerprint recognition is probably the best-known and most convenient biometric systems. Law enforcement has used this method without the support of computers for many years. With the advent of computers and high resolution electronic scanning, the method has become substantially more efficient than the original methods. As this book is written, fingerprints are generally regarded as the most convenient method of all existing biometric authentication characteristics for the end-user. However, even within this method, authentication issues or challenges remain. One key challenge is referred to as the "hidden" fingerprints. The hidden fingerprint refers to the oil imprint left on the scanning device from a previous user which creates uncertainty as to which fingerprint is being presented for verificatioin. There are scanning devices or sensors that don't work until the sensor contains the same hidden fingerprint as the current fingerprint being presented, and they use the hidden fingerprint and not the current fingerprint for verification.

The main advantage of fingerprint-based biometry is its relatively low price. Scanners are inexpensive, and their price will decrease in the future. Some manufactures produce a mouse with an integrated thumbprint scanner.

OPTICAL RECOGNITION   Eye checking, or optical biometric recognition, is considered more exact than fingerprint recognition because the eyes have more unique characteristics than the fingers. Optical recognition devices were improved during the last few years and are able to scan an eye through the lens. Error probability is about 0.00002 percent. Thus, this technology is reliable enough. However, there are disadvantages, including that people with eye illnesses, including blindness, have problems with this system.

Also, these systems cost thousands of dollars and are not too attractive for network users. At this time, there is no PDA with an integrated optical recognition authentication system, but there is a tendency to integrate digital cameras into PDAs and Smartphones, which makes it feasible that optical recognition systems will someday be used on PDAs.

Face Image Recognition   Face image recognition is not reliable yet. Such systems have been tested in airports and by police for automatic detection of criminals and terrorists. Unfortunately, these tests show that the system generates too many errors. The advantages of face image recognition are its low price and ease of use, which will be great once the system becomes reliable.

VOICE RECOGNITION   Voice recognition is reliable for telecommunication and modern computers generally by default contain the hardware support for voice recognition. Many motherboards have integrated audio chips and microphones that cost less than $10. Yet this authentication method has an error probability of 2–5 percent, which is too large for authentication. Another disadvantage is that the human voice may change during a day. Therefore, while most PDAs have voice recorder capabilities, and new PDAs have enough computing resources for voice recognition, no PDAs at this time use integrated voice recognition authentication systems.

HANDWRITING SIGN RECOGNITION   Almost all people sign documents such as postcards, checks, and official documents. This practice allows for the use of a "sign recognition" technique. Usually our signature is two-dimensional, but sign recognition treats signatures as three-dimensional. The third dimension is a dynamic characteristic of signing: the pen pressure. It makes this technique more reliable. One disadvantage of this method is that people don't sign documents exactly the same each time. Handwriting sign recognition systems are less expensive than optical recognition systems. The cost is similar to the cost of fingertip or voice recognition systems.

There are several third-party solutions for handwriting recognition and biometric authentication systems for the PDA. A well-known manufacturer is CIC Corporation, which offers a handwriting recognition system for Palm and PocketPC based devices.

RELIABILITY OF BIOMETRIC AUTHENTICATION   As we briefly discussed in Chapter 5, *"The Components of a Measured IT Response,"* biometric authentication systems are "graded" using two precise characteristics: *false acceptance rate* (FAR) and *false rejection rate* (FRR). FAR means the biometric system authenticates the user incorrectly and grants access to the user without proper rights, implying a security threat. FRR means a biometric system doesn't grant access to a person with proper rights—inconvenient for the end user but not a security risk. We must keep in mind that the published FAR and FRR results have been based on limited studies and that the measurement and algorithm technology continues to rapidly improve.

SMART CARDS   As we briefly discussed in Chapter 5, a smart card is generally a plastic card with an embedded processor chip that allows storing and processing data between users. Smart cards require a reader in order to exchange data between card and system. Introduced more than a dozen years ago, smart cards were first applied to pay phones to reduce theft. Later, smart cards were used in other areas such as electronic vaults, libraries, *global system for mobile communications* (GSM) phone identification and others. Smart cards provide tamper-proof storage of the user and account identity. They protect against security threats and careless password storage.

Types of Cards   Smart cards may have different types of embedded chip capabilities, including:

- Memory cards
  - Straight memory cards
  - Protected/segment memory cards
  - Stored value memory cards
  - CPU Microprocessor Multifunctional Cards

Also, cards may be classified according to their methods of connection to a smart card reader:

- Contact cards
- Contactless (radio) cards

MEMORY CARDS   Memory cards are not able to manage files dynamically. All communication is performed via synchronous protocols with readers. There are three primary memory card types:

Straight Memory Cards   These cards store data but do not process it. They have the lowest price for a memory unit and may be treated as data disks. They cannot identify themselves; the system utilizing them must know the type of card inserted into the reader.

Protected/Segmented Memory Cards   These cards have built-in logic to control access to memory areas inside the cards. Some may be programmed to write but protect some memory areas, while others may restrict both read and write of particular memory areas. This is usually done via system password or PIN. *Segmented memory cards* may be divided into logical segments to provide multifunctionality with each memory segment dedicated to specific card functionality.

Stored Value Memory Cards   These cards are disposable or rechargeable and are designed specifically for storing data or values. Most cards have a built-in security, which is hardwired by the manufacturer. Card memory usually acts as a decrementing counter. An example is the prepaid phone card, on which the chip has 60 memory cells. A memory cell is cleared when the appropriate time unit is used. When cleared of all memory cells, the card becomes useless and it is simply discarded.

CPU/MPU Microprocessor Multifunctional Memory Cards   Multifunctional Memory cards are the most powerful and expensive type of smart card. They have an embedded processor unit and card memory. All card resources are managed by a *card operating system* (COS). Unlike other operating systems, the COS controls access to the card memory. This capability makes the card multifunctional. For example, a multifunctional smart card may serve as a debit card and building access card at the same time. Also this technology allows upgrading the smart card's COS and built-in application to expand the card's scope of application.

Java Cards   JavaCard was introduced by Schlumberger and contains a microprocessor smart card on which the COS contains a *Java Virtual Machine* (JVM). Smart card Java support significantly simplifies application development for smart cards. Due to the smart card resource restrictions, it supports only a subset of the Java *Application Programming Interface* (API). Additionally, Java cards, as well as multifunctional smart cards, are upgradeable allowing for new applets to be uploaded into the card to extend their functionality. This upgradeability, however, creates a security threat as a result of the potential to upload malicious code or viruses.

PDA and Smart Cards   Smart cards may be used with PDAs to improve overall security. As we mentioned above, the smart card environment requires a smart card reader. PDAs with the built-in smart card reader are ready to use smart cards. At this time, the only PDA with a built-in smart card reader is HP Jornada 720/728, which supports only Gemplus 8000 smart cards.

Other manufacturers, however, produce external smart card readers for various handheld devices. These include:

- **PalmOS**   Didya.com, ACP30 Palm Pilot Smart Card reader
- **PocketPC**   Axcess Mobile Communication, Inc., Blue Jacket™ for Compaq/HP iPaq PDA series

Smart Card Standards   Several smart cards standards exist:

- ISO 7816 describes low-level interface with the smart card. It determines how data should be exchanged between the card and the reader.
- *Personal computer/smart cards* (PC/SC) describes the standard for communication with smart cards under Win32 environment. Actually *Movement for the use of Smart Cards in a Linux Environment* (MUSCLE) projects exist for porting PC/SC environment under Linux (www.linuxnet.com).
- JavaCard describes the standard for the smart card with built-in JVM.
- *Open Card Framework* (OCF) is an open standard that supports multiple platforms. It supports both PC/SC v1.0 and JavaCard standards (see www.opencard.org).

Network Connection Security

PDAs are becoming network aware. Therefore, network protocols must protect the transmitted data from interception. Secure network protocols, such as *IP Security* (IPSec), *Secure Socket Layer* (SSL), and *Tool Command Language* (TCL), are set services for authentication and data encryption. Each protocol has its own level (network level, transport level, or application level, according to the *open systems interface* [OSI] model).

Because network security is thoroughly covered in other books, is a much more mature area than on-device handheld security, and is only a recently emerging concern of PDA devices, we will provide only a quick discussion of networking-related issues here, including some key products available for the PDA market to provide transport security. Most devices either leverage a *virtual private network* (VPN)-based solution over *Transmission Control Protocol* (TCP)/*Internet Protocol* (IP), or rely on SSL as a primary method, choosing to make network-enabled applications use web browsers for information exchange. In short, the security issue around network transport must be addressed, but has ample solutions for covering *local area network* (LAN) and *wide area network* (WAN) implementations.

Internet Protocol Security (IPSec)   IPSec is an open standard for secure communication on IP networks. This protocol works on the network level in the *open standard interface* (OSI) model. IPSec is based on the standards designed by the *Internet Engineering Task Force* (IETF). It provides data authenticity, security and integrity when data is transferred via public IP networks. IPSec implements data encryption and authentication on the network level and provides complex security in the network architecture. IPSec uses the same IP packets and does not require special hardware; all existing IP network hardware is compatible with IPSec. There are requirements that IPSec-compatible systems support the *data encryption standard* (DES) encryption algorithm, but other encryption algorithms may be used:

· Triple DES
· Blowfish
· CAST
· RC5
· IDEA
· 3IDEA

Secure Socket Layer *Secure Socket Layer* (SSL) is an Internet protocol for encryption and authentication at the session level, which provides a protected channel between server and client. SSL provides server authentication and optional client authentication to avoid sniffing of the transmitted data and to prevent data modification or corruption. SSL works on the transport level and doesn't depend on application-level protocols such as *hypertext transfer protocol* (HTTP) and *file transfer protocl* (FTP). Netscape designed SSL, and SSL v3.0 is widely used at this time.

Virtual Private Network A *virutal private network* (VPN) is used to provide secure, encrypted network communications between a network and a remote host or other remote network over the public Internet. A VPN allows the establishment of an encrypted "tunnel" that protects the flow of network traffic from sniffing and interception. A VPN is based on low-level secure network protocols such as IPSec and others. IPSec-based VPN is good for fast WAN networks. Slower WANs (a satellite network, for example) may require other low-level secure protocols for decent response times.

PDAs have a variety of solutions available to utilize a VPN, using either built-in VPN technology (PocketPC) or third-party VPN solutions, such as those by PalmOS. Tables 7-10 and 7-11 list vendors that provide VPN solutions for the PalmOS (see Table 7-10) and the Pocket PC (see Table 7-11) platforms.

Data Storage Security (Data Encryption)

Data storage security is required to keep PDA data protected, preventing a hacker from extracting vital data from the device. Even if a PDA is locked, there are ways to retrieve data from the locked PDA even when the intruder doesn't know the correct password. The PDA OS may have backdoors, or the hacker may physically disassemble the PDA to get access to the internal data, as was discussed in Chapter 5. Also, if the data are stored on inserted memory cards, it will not matter whether the PDA is locked. The memory card may be simply removed from the

TABLE 7-10.   PalmOS VPN Providers

COMPANY	PRODUCT NAME
Merqic, Inc.	Merqic VPN
Certicom Corporation	Movian VPN

TABLE 7-11.   Pocket PC VPN Providers

COMPANY	PRODUCT NAME
Certicom Corporation	Movian VPN
Check Point Software Technologies Ltd.	VPN-1 Secure Client
Columbitech	Wireless VPN client
Entrust, Inc.	Web Portal portfolio
Epiphan Consulting Inc.	VPN and LinkSpy
Funk Software, Inc.	AdmitOne VPN
Linksys Group Inc.	Instant Broadband EtherFast Cable/DSL VPN Router built-in IPSec coprocessor
Symbol Technologies, Inc.	AirBEAM Safe wireless VPN
V-ONE Corp.	SmartPass client

card slot, and the data are readily available. The solution is to encrypt critical and private data on the device-managed storage media.

As illustrated in Chapter 6, PDA Defense includes both user authentication and data storage security for built-in storage in RAM and external storages such as memory cards, Compact Flash, *Secure Digital* (SD)/*Multimedia Card* (MMC), and Memory Stick.

To ensure that the data are both adequately secured and—this is important—also reliably accessible, well-known encryption algorithms must be used. There are several types of encryption algorithms: block ciphers and stream ciphers, symmetric ciphers and asymmetric ciphers.

Which cipher algorithms works best for a PDA? PDAs, in contrast to desktop computers, have substantially limited power, storage capacity (*random access memory* [RAM]), data rate transfer (bus) and CPU speeds. Block cipher produces ciphered data with the same length as the source data; stream ciphers produce data with increased length. Thus, if we decide to use stream ciphers, we will need additional storage for ciphered data, equal to or larger than the storage space for the original plain data. In addition, block ciphers are faster than stream ciphers. Table 7-12 provides a brief discussion of several block ciphers to be considered for preferred use.

When selecting or working with an encryption product, the configuration will help determine which algorithm will be used. As administrators of security policy, we have the opportunity to make cipher decisions that affect both performance and strength of our PDA security approach. When we get to section called "Why Cryptography?" we will take an even more detailed look at the various algorithms.

How Locked Devices Resist Intruder Penetration

The PDA may be directly attacked with the intent of pulling the data straight from the device. We will discuss the implications

TABLE 7-12.   Cipher Examples

DES	Key length is 56 bits. This algorithm was standard for a long time, but it is now compromised. More reliable algorithms should be used.
Advanced Encryption Standard (AES)	A new encryption standard recommended by the *National Institute of Standards and Technology* (NIST) for commercial and governmental use.
Blowfish	A well-known, reliable block cipher, introduced by Bruce Schneier. Blowfish is resistant enough for cryptanalysis.
XOR	A simple, fast algorithm. Unfortunately, it is unreliable for cryptanalysis and cannot be recommended for protecting data. Nevertheless, it can be used in combination with other more reliable block ciphers.

of attack when the device is locked and requires a password for access. Specific scenarios include:

- The locked device doesn't have a network or other connection, excluding the keyboard, physical interface ports and touchpad.
- The locked device may have network connections or other low-level connections, such as *Infrared Data Association* (IrDA) or LAN.

Connectionless Devices   Connectionless devices are not network aware, and they have no input methods except for the keyboard, physical interface ports, or touchpad. The only way to turn off the "lock" mode is to enter the correct password on the touchpad or keyboard or to simulate such through interface ports. The intruder may penetrate a connectionless device if he or she knows about certain OS backdoors. Some early OSs had a built-in debugger that could be activated by special key codes, creating a connection with a remote desktop. Another way to get information from the connectionless device is by physically removing the memory chips and copying the information to external storage.

The device resists penetration access passively or actively. Built-in security provides passive resistance; it rejects incorrect passwords but takes no further action. The active method means that the PDA detects attacks and responds with various actions, which may include destroying the data on the PDA. For example, if too many incorrect passwords are attempted, the data might be erased.

Connected Device   The connected device has network connections or other low-level connections such as IrDA or LANin addition to the touch screen and optional keyboard, resulting in additional vulnerabilities. The PDA may have several types of connections, including IrDA, Bluetooth, LAN, cellular, and wireless connections (802.11b or *Wireless Fidelity* [Wi-Fi]). Hackers familiar with the network protocol and how specific devices use the protocol may know how to send unauthorized programs to the device. Many devices, if not adequately locked down, may have libraries and background processes that would respond to incoming network activity, such as a request to install new software that could subsequently disable the security on the device.

If a network-originated attack gains access to execute code on the device behind the locked user interface, the security software will be compromised. Any security software should ensure that all network connections on the device are disabled while the device is locked.

The role of PDAs has changed, however, so that the once-simple operation of the device, which executed code only as a response to user input, has expanded. The PDA increasingly looks more like a desktop computer, executing various background programs with administrative privilege while the device is locked. The concept of a locked device no longer means "no execution of code allowed," but "no user input allowed." The user may want certain programs such as cellular phones to work while the device is off. Security software authors have the challenging task of creating software that allows network connections to operate only in authorized fashion when the device is locked. This is not a simple problem to resolve and, in many cases, requires the cooperation of the software company creating the wireless application.

Why Cryptography?

Everybody has personal information they wish to keep confidential: *Social Security Number* (SSN), credit card numbers, bank account numbers, PIN codes, friends' phone numbers. Financial software may fill your PDA with your personal valuable information, much like a bank vault. For example, PocketPC devices have built-in Microsoft Money software, and there is plenty of third-party financial software, as well. PDAs often contain your work data, such as customer or prospect lists, sales volumes, pricing data, and departmental notes.

While user-owned PDAs contain personal information, company PDAs may contain data that, if lost, could be devastating to the company: results of investigations and developments, technical information, strategic plans and other such documents. Healthcare companies may keep patient health information on the PDA, and this information must be encrypted according to existing federal law as enacted in *Health Insurance Portability and Accountability Act* (HIPAA).

In the desktop environment, most security strategies have essentially accepted these devices' data storage as being at risk, and moved data storage to the network, resulting in the desktop's becoming an authenticating device to gain access to secure network storage.

However, the desktop model is a more sophisticated security environment than current PDAs at the OS level in that it provides object-level permissions and role-based user accounts that the OS uses when the user tries to open a file. The OS retrieves user permissions and file permissions, and compares the set of permissions. If a user's permissions are sufficient, the OS allows the file to open; otherwise, the file access is denied. Usually, access rights are assigned and controlled by a system administrator or supervisor.

In part, because of the focus on access rights and a number of other issues, including the level of background processing and system-level cakks to data sources, most desktop environments make little use of encryption as a secondary protection strategy in the event object and user-level authentication properties are bypassed.

The PDA OS works with more limited security concepts than the desktop OS. PDAs do not recognize multiple users with differing access rights. The person who has access to the device has access to everything on the device, that is, all or nothing. Some exceptions exist; Embedded Windows NT and Embedded Windows XP have full support of access rights and *Encrypted File System* (EFS) support. But a shortage of PDA devices with embedded WinXP exists; manufacturers usually use Windows CE.

If your PDA has settings for network connections to your company's network, your network settings may be retrieved from a stolen device and the hacker may get access to the company's network. A stolen PDA can become a nightmare for system administrators, as we discussed in detail in Section 2. The following provides a more detailed look at the various cryptography algorithms introduced in Table 7-12 when we began our discussion of data storage.

XOR XOR means "exclusive OR," which is a bit-oriented operation. First, let's describe the operation OR: Take two bits. If one bit OR another one is 1, the result will be 1. Otherwise, the result is 0. This example is illustrated in Table 7-13.

Exclusive OR may be described as follows: Take two bits. If *only* one of these bits is 1, the result is 1. Otherwise, the result is 0. This example is illustrated in Table 7-14.

XOR is a useful operation in cryptography; if one bit belongs to plain text and another bit belongs to the key, sometimes the resulting bit is changed, but sometimes it is not. In the following example, the first string (0x45268456) is the data being encrypted, and the second string (0x8D1E6A65) is the key. The resulting XOR value (0xC838EE33) is the encrypted data, as illustrated in Table 7-15.

The result of the operation will be cipher text. If we perform the XOR operation with the original key and cipher text, the result will be the original text (0xC838EE33 XOR 0x8D1E6A65 = 0x45268456). XOR is popular in cryptography because the

TABLE 7-13. Explanation of Simple Rule or Algorithm for Executing an OR Operation

0 OR 0 = 0
0 OR 1 = 1
1 OR 0 = 1
1 OR 1 = 1

TABLE 7-14. Explanation of Simple Rule or Algorithm for Executing an XOR Operation

0 XOR 0 = 0
0 XOR 1 = 1
1 XOR 0 = 1
1 XOR 1 = 0

TABLE 7-15. Illustration of the Resulting Binary String Produced by the Combination of an Original String with a Similar Length Key String

0x45268456	=	0100 0101 0010 0110 1000 0100 0101 0110
0x8D1E6A65	=	1000 1101 0001 1110 0110 1010 0110 0101
0xC838EE33	=	1100 1000 0011 1000 1110 1110 0011 0011

second XOR operation with the same argument produces the original number.

Digital Encryption Standard   IBM developers, under the direction of Horst Feistel, developed an algorithm named Lucifer. They worked with the *National Security Agency* (NSA), and the result of that work is the DES algorithm.

DES is a block cipher, which uses a 56-bit key for building a key table. DES performs bit operations on the plaintext using the key table. To recover plain text from the cipher text, operations must be performed in reverse order. DES became available to the public and was investigated thoroughly. In the 1980s, cryptographers decided that DES didn't have a weakness. Even the fastest computers of the time required several years to crack one ciphered message.

In the 1990s, cryptographers decided that DES could not be the encryption standard anymore. Computer performance had increased significantly, and it became possible to successfully crack the 56-bit key using a brute-force attack. Also, crypto-analysts investigated all potential weaknesses, and they decided that an algorithm crack is possible.

At the RSA conference, Electronic Frontier Foundation cracked the DES key in less than 24 hours. Now DES cannot be treated as encryption standard.

Triple DES   One of the DES replacements is Triple DES. The name described the algorithm itself. Triple DES executes DES three times. Each key length is 56 bits so the total key length is 168 bits. However, a weakness was found in the Triple DES algorithm, making it less than perfect for encryption. The total strength of Triple DES is equivalent to the block cipher with the 108-bit key.

Blowfish   Bruce Schneier invented the blowfish encryption algorithm. It is sufficiently fast and doesn't require a lot of resources. Blowfish is 64-bit block cipher. Unlike Triple-DES, it contains no known weaknesses.

Advance Encryption Standard   The NIST announced a plan on January 2, 1997, to choose crypto algorithm from among 15 candidates for its new standard. Specialists analyzed every detail of the algorithm. The criteria for the new algorithm were the absence of any weakness, high performance (i.e., reasonably fast) and low required resources. Most of the 15 algorithms presented were too slow or consumed too many resources. On October 2, 2000, NIST announced the winner, Rijndael, which was developed by two Belgian scientists: Vincent Rijmen and Joan Daemen. *Advance Encryption Standand* (AES) may use 128-, 192-, and 256-bit keys, and is recommended by NIST for government and Enterprise use.

NIST comparisons illustrate how AES is better than DES. If a special hardware-based DES cracker can crack a DES key in one second, the same hardware will spend 149 trillion years to crack the 128-bit AES key. We consider this difference meaningful!

Access to Device Storage Bypassing the OS

One of the PDA OS functions is organizing data storage. The file system is only one data storage type. PalmOS has its own non-file-based data storage, and PalmOS v4.0 and higher support an external file system for memory cards only. WinCE supports several data storages types as well, including:

- File system storage.
- System registry storage.
- Database system storage.
- Kernel internal storage that cannot be accessed by application.

From a technical point of view, each storage type is a binary array in which some areas are designated for storage objects names, objects attributes, allocation tables, and so on. Usually, the OS handles requests to and from storage, as illustrated in Figure 7-4. But the data inside the storage may also be accessed directly. In this case, internal system data structures should be processed by special software, as illustrated in Figure 7-5. This technique is called *data recovery*.

Data recovery is a powerful industry, but no data recovery software for PDAs is available to the public, even though it does exist. For example, PwC Consulting, now part of IBM Global Services, has special software to recover data from PDAs and mobile phones for police and other organizations. At this point we will not review physical access to media, as was discussed in Chapter 5, except to say that our encryption discussion primarily addresses situations in which the device OS has been compromised, leaving encryption in place but allowing access to the storage media. See Figure 7-4.

POCKET PC OS PLATFORM

PocketPC-based devices use the WinCE core and support security providing user access authentication and network connection security (SSL, for example) including VPN support.

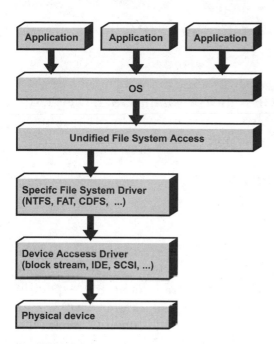

FIGURE 7-4.   OS requests to and from storage.

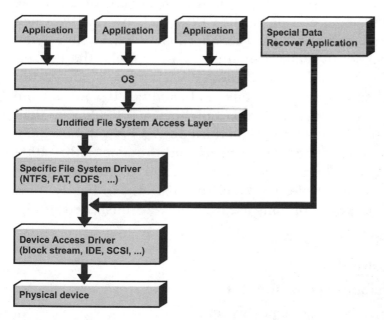

FIGURE 7-5.   Internal data structures.

What Happened to WinCE, PocketPC 2000, 2002, and .Net?

There are many WinCE-based devices, many of which have a unique look and feel. WinCE is the core of the OS; however, the shell that surrounds it is also important. Microsoft produced several shells based on WinCE 3.0:

- H/PC 2000 (handheld PC)
- PocketPC
- PocketPC 2002

Microsoft recently released WinCE.NET (which has a v4.0 version number) but, as of this writing, there is no PDA-specific shell. Microsoft has promised to release one in 2003, but until it does, only the standard Win95-like shell is available. Some manufactures release devices with the actual version of WinCE.NET.

What does "version of shell" actually mean? If you know the type of shell you are using on the PDA, you know exactly which components exist on that PDA because each shell version has its own set of security modules. For example, the PocketPC 2002 shell has a built-in VPN support module, but PocketPC shell doesn't contain that module. If you have an old PocketPC device and you want to use VPN on it, then you need to find an appropriate third-party VPN solution.

Where is the built-in security on the PocketPC 2002 devices? It is on the Settings panel, under a Password icon, containing a Password settings panel, as illustrated in Figure 7-6.

Three password options are available, as illustrated in Figure 7-7:

- No password.
- Simple four-digit password (PIN).
- Strong alphanumeric password.

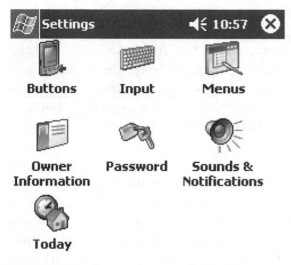

FIGURE 7-6.   Password icon on Pocket PC.

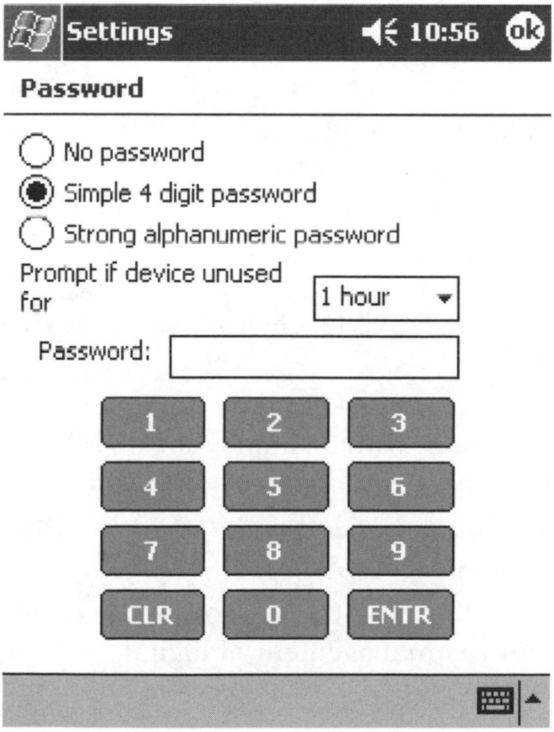

FIGURE 7-7.   Password options on Pocket PC.

The oldest PocketPC devices offered a four-digit password (PIN) only. PocketPC 2002-based devices offer the use of strong alphanumeric passwords. In this case, you will be prompted to enter a password that is no fewer than seven characters, contains both upper- and lowercase characters, and has at least one special character. Also, you can specify when the device locks, either immediately after switch-on or after a specific period up to 24 hours.

One of the Windows CE features that applications cannot handle well is the device power-off event; the built-in security application—as well as all other applications—is able to handle the device power-on event well but power-off overrides application efforts to cleanly shutdown. As a result, if a security application for example specifies that the device should lock as soon as the device powers off, in reality the shut down of the device will occur prior to the security application having the opportunity to execute the lock commands. As a result, the device will be locked only after it is powered back on and the security program recovers and executes the locking routine.

As illustrated in Figures 7-8 and 7-9, a locked device requires a PIN or password to be entered to access all the device resources, depending on the type of password selected.

Strengths and Weaknesses

The PocketPC (old, 2002, and .NET) environment has rich security support, including:

- Device access authentication, with a simple four-digit PIN and strong alphanumeric password for PocketPC 2002.
  - *Cryptography application programming interface* (CAPI) support, as illustrated in Figure 7-10.
  - Digital certificate handling. WinCE supports a subset of CAPI v2.0 to support the management of digital certificates.

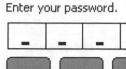

FIGURE 7-8.   PIN entry screen.

FIGURE 7-9.   Strong alphanumeric password entry.

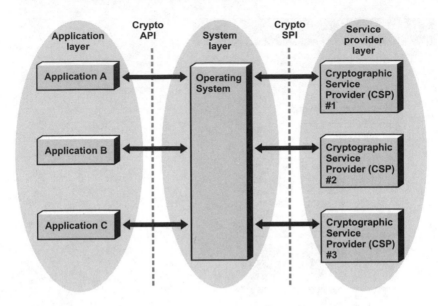

FIGURE 7-10.   CAPI Architecture Illustrated.

- Network security support. *Security Support Provider Interface* (SSPI) provides a common interface between network applications and security provider modules, as follows:
  - Windows NT *LAN manager* (NTLM).
  - SSL v2.0 ad v3.0.
  - *Private Communication Technology* (PCT) v1.0.
  - Smart card support.

One threat develops from the powerful nature in which the device creates a TCP/IP connection between desktop and the locked device allows for brute-force attacks to be more easily attempted. The following text examines first a connectionless device and then a connected or network-enabled device to evaluate security issues.

Security on the Connectionless Device   The connectionless device is protected only by built-in authentication to access the locked device. As we mentioned earlier, the locked device can be accessed via ActiveSync if the user knows the correct password. ActiveSync displays a prompt dialog on the desktop when the user inserts the PocketPC device into the cradle, and the user enters the PDA password in this dialog to establish a connection with the PDA. No restriction exists on the number of attempts a user can enter a password, so a brute-force attack is possible.

Also, a significant issue arises with the remote password. If the PDA has a system password, ActiveSync requires that the password be entered each time the PDA is connected via the cradle. However, ActiveSync has the option to save the PDA password on the desktop, and ActiveSync uses this password automatically when PocketPC is connected to the desktop via cradle. The password is stored on the desktop in scrambled form and can be recovered easily by anyone who has access to the desktop.

A connectionless device has no network connection; the only way to get information from the locked PocketPC device is to bypass the activated lockout screen of either the built-in security or the third-party security solution. If the PocketPC device is locked, the lockout screen with the password prompt is active. At this point, the user must enter the correct password to close the lockout, allowing access to the PocketPC. Pocket PC-based devices have a lockout screen similar to the digital lock keypad with screen buttons that are labeled by digits 0 to 9. It is easy to use but not very secure. The new PocketPC 2002 device has an advanced lockout screen that prompts for a more secure alphanumeric password.

On the oldest PocketPC devices, PDA Defense solves this issue by requiring the user to enter a strong alphanumeric password. To eliminate the possibility of a brute-force attack on the password, the number of attempts to enter a password may be restricted, and the ability to establish ActiveSync with the locked device is eliminated. The ability to establish ActiveSync connection with the locked device is a serious security flaw because the ActiveSync connection runs various debug tools and simply transfers all the available information from the locked device, including files and databases, from RAM storage.

Every PocketPC device is able to store information on the external memory storage—memory cards in several available formats, such as Compact Flash or SD/MMC. Some devices have a built-in memory card slot. For those that don't, special expansion devices such as jackets give the PocketPC device access to the use of external memory cards. For example, Compaq iPAQ 36xx and 37xx series can use jackets with memory card slots, and Compaq iPAQ 38xx and 39xx series have a built-in Secure Digital/MMC card memory slot.

PocketPC devices store information on the external memory cards in the usual unprotected form, and the memory card may be removed from the slot with ease. Thus, we cannot store vital information on the external memory cards in order to avoid unauthorized access to the data on those cards.

PDA Defense solves this issue, too. It allows for the creation of protected virtual memory cards that use an external memory card as the underlying media. Each virtual protected card has its own password and encryption key, and the user should enter the correct password to use the protected card. If the user's password is correct, then PDA Defense recovers the encryption key and mounts the protected card on an additional folder under the root directory in the system. All read/write operations are encrypted/decrypted, and the information inside the virtual protected card is always encrypted. Conversely, applications work with files on the virtual protected card as they do with normal files, and the protection layer is completely transparent for end-users.

Security as Network Connectivity and Peripherals Are Added   It is becoming increasingly less likely that the PDAs in your organization will be connectionless PDAs. For example, a user can connect to the Enterprise's LAN using a network card or a wireless card. He or she could also connect to the Internet via a modem or an Enterprise Internet connection. This increases the risk that private information may be exposed or that the

PDA may be attacked from outside, which is a serious issue if the PDA contains confidential or commercial information.

Each network connection opens a conduit for transmitting data to the outside world, and you cannot guarantee that your data will not be stolen on its way to your recipient! Thus the usual unprotected network connection should not be used to deliver confidential information via public Internet network. Even private dedicated networks should not be fully trusted for data transfer because wireless elements at or near the end points may compromise your otherwise private network.

The solution to this issue is VPN. A VPN contains a Virtual channel, and all data inside the channel are encrypted using session keys. Only members of the VPN can access and read the real data. PocketPC devices use third-party solutions for VPN, but PocketPC 2002 and .NET-based devices have built-in VPN support. Make sure that your network or security group allows tunneling protocols through your corporate firewall, or see if you can leverage the any pre-existing VPNs already supported in your organization. VPN access conveniently establishes a secure connection through your existing network infrastructure, and it may reduce your overall PDA security costs significantly.

Web-based Internet services require a protected connection. Usually SSL is used for protected web services. All PocketPC-based devices have SSL, PCT support. Modern PocketPC 2002 devices (iPAQ 39xx series for example) have Bluetooth and wireless connections via third-party wireless cards (some PocketPC 2002 devices have built-in wireless connectivity). However administrators must carefully control the configurations on the PDAs of this wireless connection; otherwise wireless connectivity may become the new security hole in Enterprise networks.

PALMSOURCE OS PLATFORM

As illustrated in Figure 7-11, PalmOS has security support through built-in device access authentication. PalmOS-based devices utilize a password to lock and unlock the device. It also supports private and public records. The user can assign a "pub-

FIGURE 7-11.   Palm security application.

lic" or "private" attribute to individual records and change mode
to display private records. All PalmOS versions support the display or hiding of private records as displayed in Figure 7-12; PalmOS v3.5 and higher versions support masking of private
records. In the "mask" mode, applications show the presence of
private records, but the actual content of private records is not
shown; the user must switch PalmOS to the "show" mode. Protecting information using private records is not reliable, however, because PalmOS does not protect the actual contents of
the private records and because the only difference between
public and private records is a special attribute flag that is handled by the application programmatically. Users should not mis-
take the use of private records as real protection.

Because the PalmOS does not have built-in support of
secure network connections, the user should use third-party
solutions to protect the connections. PalmOS built-in security
is available from standard launcher for all devices with PalmOS
3.0—PalmOS 4.1 and SONY NX series with PalmOS 5.0. Palm,
with its Tungsten series based on PalmOS 5.0, moved built-in
security from the launcher screen into the Preferences screen.

PalmOS 4.0 built-in security allows locking the device automatically after specified conditions as follows and as illustrated
in Figure 7-13:

FIGURE 7-12.   Palm security settings for show mode: show, private, and mask.

FIGURE 7-13.

·   Device power-off.
·   At a preset time.
·   After a specified period of device inactivity.

Palm 2.0, 3.x, 4.x, and 5.x

PalmOS has a long development story. The first successful Pal-mOS version was v2.0. Then Palm, Inc. released v3.0, v3.1, v3.5, v4.0, and finally v5.0, which were improved versions with only a few changes from v2.0. Palm improved the OS in general and added some APIs.

For example, PalmOS v4.0 supports a file system for external memory cards called *Virtual File System* (VFS). Palm also introduced phone API support and many other APIs that enable use of mobile phones with PalmOS-based devices for wireless Internet access. But the latest PalmOS, v5.0, is a new step in the PalmOS history. In contrast to Microsoft Windows CE, the original PalmOS was developed for a specific microprocessor, the Motorola Dragonball. Motorola released new versions of the processor, but the core processor system has not changed.

At some point, Palm recognized that a slow core processor decreases the competitiveness of the PalmOS in comparison with WinCE. Thus, the core processor has been changed to the ARM-based CPU in PalmOS 5.0. The core was rewritten from scratch and is executed on the *asynchronous response mode* (ARM)-based CPU. To allow for the use of earlier applications created for PalmOS v3.0 and 4.0, PalmOS v5.0 has a special emulation environment called *Palm Application Compatibility Environment* (PACE), as illustrated in Figure 7-14.

All existing applications for the Dragonball processor are executed under PACE, and almost all (80 to 90 percent) "well-behaved" applications execute normally, as long as they don't use a special nonstandard technique. PACE allows for the migration to PalmOS 5.0 without pain and allows existing applications to be adopted by the new PalmOS without recreating the applications. Dragonball applications almost always execute faster under PalmOS 5.0. PACE is an emulation environment, and although the ARM CPU is faster than the Motorola Dragonball, the emulation requires more CPU time than the execution of the "native" code.

Applications for PalmOS 5.0 can use the advantages of the new fast CPU directly. Developers can decide what code is time-critical and recompile code fragments for the ARM CPU. This is accomplished by calling special OS routines and specifying code fragments in ARM native code. The code is executed by the system in full speed (versus emulation), and overall

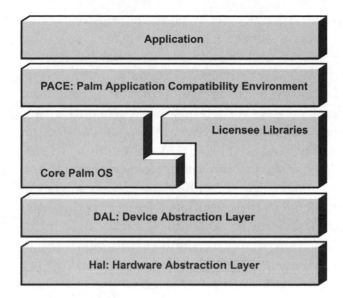

FIGURE 7-14.

performance should increase significantly. Palm, Inc. points out that all *graphical user interfaces* (GUIs) and graphic OS routines are ARM native code and that applications should not try to optimize them.

Strengths and Weaknesses

PalmOS v5.0 has an improved security support system. It contains a built-in security application that allows the user to lock the device at a specific time or after a specified period of inactivity. Palm Source announced that v5.1 introduces a new level of security support, Crypto Manager, which provides unified encryption module support and SSL support.

Security on the Connectionless Device   Connectionless PalmOS PDAs have password-level authentication. A user may assign a system password and lock the device to deny others access. Unfortunately, the PalmOS built-in security has a serious security flaw in versions prior to v4.0. PalmOS uses its own proprietary algorithm to store a scrambled password on the device; a web site called @stake (www.astake.com) investigated and published the "scrambling."

Also, PalmOS v3.5.2 and earlier versions have a special debug mode that allowed direct access to the data on the device, even while locked, allowing for the retrieval of vital information, including the scrambled password. Thus it is a straightforward exercise to obtain the scrambled password via IrDA connection if the Palm device HotSync is set to IrDA connection. In this case, another device can easily simulate a HotSync connection via IrDA; HotSync transfers the scrambled password via IrDA to the Palm Desktop, which in turn prompts the user on the desktop for the correct password. In addition, a hacker may introduce a rogue application that unscrambles the password to recover the original password. Note that these security flaws are corrected in the next versions of PalmOS: v4.x and v5.x.

The latest PalmOS versions store just the password hash. At this time PalmOS uses the MD5 algorithm to generate and store the password hash on the device. PalmOS has another problem: outside applications may be executed on the PDA if they are received via the beaming (IrDA) process. After receiving an application, the PalmOS asks whether the user wants to accept the received application; if the user agrees, PalmOS makes the application permanent and executes it simulating the HotSync process. This works in like manner to applications installed via a normal HotSync operation.

We also must consider security on external media. PalmOS v4.0 and newer devices are able to store information on the external memory cards in varying formats (Secure Digital/ MMC, Memory Stick). Also, there are devices with PalmOS 3.5 and Compact Flash memory card support: TRGPro (discontinued) and Handera 330 by Handera. All Palm-manufactured devices with PalmOS 4.0 or higher have an integrated SD/ MMC memory card slot. Palm-compatible devices manufactured by SONY or Acer use Memory Stick memory cards.

By default, PalmOS devices store information on external memory cards in an unprotected form. Of course, the memory card may be removed from its slot with ease. Thus, without additional security, we cannot store vital information on the external memory cards because they can easily fall into the wrong hands. As illustrated in Chapter 6, PDA Defense, like other security solutions, solves this problem by creating protected virtual memory cards that use the external memory card as underlying media.

Each virtual protected card has its own password and encryption key that the user must enter to access the protected card. If the user's password is correct, PDA Defense recovers the encryption key and mounts the protected card as an additional folder under the system root directory.

All read/write operations are encrypted/decrypted on the fly, meaning that data on the virtual protected card are always encrypted. Applications work with files on the virtual protected card as they would with standard files; the protection layer is completely transparent for all Palm applications.

PDA Defense implements a different virtual protected card support method for PalmOS 4.x and PalmOS 5.x because of differences between these PalmOS versions. PDA Defense for PalmOS 4.x catches all requests to the VFS API and encrypts/decrypts data on the fly as illustrated in the process map in Figure 7-15. Thus the user works with files on the protected card transparently, and PDA Defense protects files under folder "/PALM" only for compatibility purposes. PDA Defense for PalmOS 5.x uses another technique; it works like the PocketPC version of PDA Defense, creating a virtual card utilizing the external memory card as underlying media, as illustrated in the process map in Figure 7-16.

Security as You Add Network Connectivity and Peripherals
As discussed earlier, network connectivity and peripheral devices add to the vulnerability of PDAs, including PalmOS-based devices. For example, removing the Springboard module from the Handspring Visor Deluxe breaks the standard lock mode. Newer PalmOS-based devices also have wireless connectivity. For example, Palm Tungsten-T has integrated Bluetooth. Handspring Treo and Tungsten-W have GSM (cell) capabilities, allowing for a wireless Internet connection. Palm i705 has the ability to connect via the Palm.net Internet service, which allows for a permanent

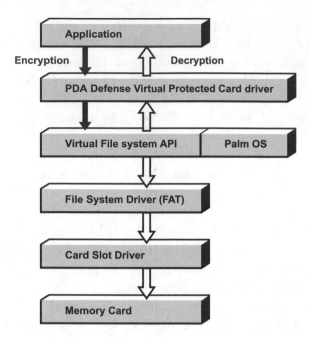

FIGURE 7-15. Interception for encryption/decryption.

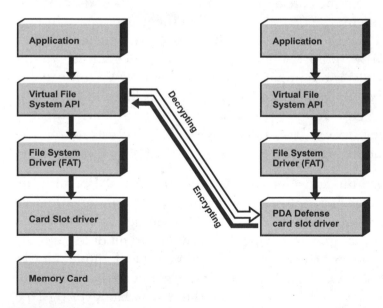

FIGURE 7-16. Virtual card as approach encryption/decryption.

Internet connection that is non-secure. PalmOS (at least v5.0) does not have built-in support of secure network connections; third-party solutions should be used to secure any network connections from the device.

For Enterprises, the system administrator should have the ability to control network settings. Without this control, it is possible to have undesired access from PDA devices to the Enterprise network as a result of wireless-enabled PalmOS-based devices. Also, any IrDA security flaw should be eliminated using third-party solutions. For example, PDA Defense Enterprise has the option to disable IrDA receiving permanently as controlled by the system's administrator.

RIM, SHARP, NOKIA, AND OTHERS

So far we have discussed PocketPC and Palm-based devices. Other devices are on the market, as shown here:

- RIM with RIM OS.
- Linux-based PDA (Sharp Zaurus and others).
- Java-based devices, including a new RIM product.
- Symbian OS-based SmartPhones, including Nokia Smartphones (9210, 60 series).

These devices contain differing levels of security support, but all have a lock mode with password authentication. Network security support is different for each device type.

RIM: The Concerns You Should Have

The Canadian company *Research in Motion* (RIM) manufactures the RIM device. Its strength is organizer functionality coupled with wireless email. The RIM PDA is based on the RIM

TABLE 7-16.   Summary List of Various RIM Products

RIM 850
RIM 950
RIM 857
RIM 957
BlackBerry 5810
BlackBerry 5820
BlackBerry 6510
BlackBerry 6710
BlackBerry 6750

OS). Several models are available, which were discussed in chapter 3 and are displayed quickly here in Table 7-16:

Models RIM 850 to 957 are based on the RIM OS 2.1 and work in the DataTAC or Mobitex data networks. These devices transfer data, and the only available wireless application is email. Additional services that expand RIM devices capabilities include support of HTTP and other Internet protocols. These devices use simple security models. The user may assign a system password and enable a lock mode. The device keeps the wireless connection in a locked mode as well. Desktop software prompts the user to enter a password if the user inserts a RIM PDA into the cradle.

Models RIM 5810 to 6750 are newer devices based upon the updated RIM OS with Java Virtual Machine and with support of *Java 2 Micro Edition* (J2ME). This feature allows execution of all J2ME compatible applications and makes the RIM platform more flexible and appealing for application developers. These new models work in the GSM/*general packet radio service* (GPRS) network and have fully functional TCP/IP support, supporting Internet browsers. With the new key feature—voice support using the GSM network—these devices may be regarded as Smartphones. These models have built-in security similar to the previous RIM models.

RIM uses secure network connections for delivering email between the device and the BlackBerry network gate using crypto modules by Certicom. But keep in mind that email is still unsecured on the way from the Enterprise or *Internet service provider* (ISP) email server to the BlackBerry gate and back. Additionally, RIM devices have a serious security issue. All information is stored in the FLASH memory without encryption, and the RIM *software development kit* (SDK) has a special tool that retrieves all Flash-stored information, even from a locked device.

Thus, built-in security doesn't provide complete security. PDA Defense for RIM is available for RIM models 850 to 957. It can be used to encrypt *personal information manager* (PIM) and email data on the locked device and resists attempts to retrieve FLASH-stored information bypassing the OS.

Symbian: Nokia and European Smartphones

There are several new Smartphone devices that combine PDA functionality and GSM mobile phone functionality. These devices are based on PocketPC 2002 Phone Edition or Symbian OS. Some of the devices were described in chapter 3 and are further illustrated in Table 7-17.

These devices are equipped with Symbian OS 6.0; they are smart devices with PDA functionality and with an integrated GSM cell phone. They use OS standard built-in security and *Subscriber Informatin Module* (SIM) lock functionality.

What is SIM lock? Each GSM mobile phone has a special chip card, SIM. SIM allows the user to store a four-digit user

TABLE 7-17.   Summary List of Various Symbian Based Products

Nokia 9210 series
Nokia 7650
Nokia 60 series, 3650 (available in 2003)
Sony Ericsson P800 (available in 2003)

PIN; if the device is locked, it prompts the user to enter a PIN to use the mobile phone. If the user exceeds a maximum number of attempts (usually three), the PIN is disabled and the user must enter a longer *Personal Unblocking Key* (PUK) code which is the 8-digit number you get on your SIM card certificate when you buy your phone. If the user exceeds the maximum number of attempts to enter the PUK code, the device is locked "forever." Actually, it means that you need to visit a service location to unlock the device.

There is a problem applying data encryption on Smartphones. For example, when the phone receives a call, it will determine the caller number with caller ID and look up the number in the address book for display. If the address book is encrypted and the device is locked, decrypting the address book is impossible in the time frame of the call. This is a common problem with Smartphones when data access is required on the locked device.

Linux, Java, and the Rest

Several Linux- and Java-based devices are available. Some existing devices, Compaq iPAQ and Palm V for example, may be modified to use embedded Linux instead of the original OS (WinCE and PalmOS). But several native Linux-base devices exist, as illustrated in Table 7-18, and discussed briefly in Chapter 3, *"The Power Resource Guide to Understanding Where Security Must Be Achieved."*

Not all these devices are available on the market, and some are available only in Asian markets. At this time, Linux-based PDAs occupy only a small segment of the PDA market, but they have great potential in terms of security. Because Linux is an open-source operating system that can be modified by anyone, a lot of security support modules are available, such as fingerprint scanner support and encrypted file system support. For example, the CDL Paron secure PDA has a built-in fingerprint scanner, hardware random number generator, hardware encryption accelerator, and built-in support for several encryption algorithms (DES, Triple DES, and RC4). Java-enabled PDAs also have the capability of encryption and message digest support, according to Sun Java specifications.

TABLE 7-18.   Summary List of Various Linux-Based Products

GSPDA V-2002 (for Chinese market)

Sharp Zaurus SL-C700, SL-A300 (for Japanese market)

Sharp Zaurus SL-5500, 5600

CDL Paron secure PDA

Informant Kaii

Invair Filewalker

Royal Lin@x

PowerPlay III, V by Empower Technologies

Softfield VR3

G.Mate Yopy

FIGURE 7-17.   CDL Paron secure PDA.

CONCLUSIONS FOR PASSWORDS, ENCRYPTION, AND THE OS

After our detailed discussions of passwords and encryption, and our review of the operating systems, we can now discuss practical issues of attacks on PDAs and how to mitigate them. If you have survived the Password discussion and were able to follow our review of the XOR encryption process, you now understand the strengths of varying passwords and the implications of various encryption algorithms. If you enjoyed that, Chapter 8, *"White Hat Hacking Threats and Mitigations,"* will not seem overly technical. While the preceding discussions were detailed, they helped you to understand the key topics of security and how to approach managing risk as you move forward.

CHAPTER EIGHT

HACKING THREATS
AND MITIGATIONS

The *personal digital assistant* (PDA) stores and manages a
wide range of information, including notes, regular email,
personal contacts, calendar info, and an increasingly large number of applications that can be integrated with Enterprise *customer relationship management* (CRM) and *Enterprise resource
planning* (ERP) systems. In addition, PDAs are increasingly
using a wide range of wireless networking technologies for receiving and transmitting this data over private and public networks. Examples of these uses can range from physicians
keeping private patient information to military units storing battle plans. With the increasing level of sensitive information
being stored, managed, and transmitted from these devices,
PDA hacking has become a real threat to be mitigated by *information technology* (IT) organizations worldwide. Threats to this
information may come in many forms:

· Lost or stolen PDAs, allowing a series of short-range
  interfaces to be used by intruders as access points to
  attempt and compromise stored information on the device,
  such as a keyboard, *universal serial bus* (USB) port, or
  infrared (IR) device.

· In-transit interceptions of information being transmitted by
  PDAs over public or private networks between the PDA and
  the user desktop or Enterprise *local area network*
  (LAN)/*wide area network* (WAN).

THREATS: EXPLOITING MULTIPLE INTERFACES

Increasingly, PDA devices provide a variety of different interfaces or channels through which they can be accessed, controlled, and generally used to exchange data with the outer world (desktop, another PDA, LAN, or global Internet/WAN). All these channels can be used to intercept or retrieve data from the PDA. The general classes of channels available include the following:

- External keyboard.
- Infrared port (*Infrared Data Association* [IrDA]).
- Radio frequency wireless networks (Bluetooth, 802.11b, and so on).
- Different extension ports and memory card slots.

PDA and Physical Proximity

The ease with which devices are stolen and the frequency with which devices are lost clearly make short-range interfaces an important threat to understand and manage. With physical access to the device, a wide range of options become available for compromising the device's security measures. The following discussion explores some of these access approaches with the *operating system* (OS), followed by a specific discussion that further details how threats can occur.

External Keyboards: Widely Used and Supported   Most modern PDAs provide either no keyboard or limited thumb-based keyboards due to size and weight constraints, but an external keyboard can be connected to almost all PDAs. Although the main purpose of an external keyboard is to make entering large amounts of text easy, the keyboard can also be used to enter passwords when the device is locked. In a brute-force attempt to compromise a password, the external keyboard can be useful.

The following scenario provides an illustration. Let's say a locked device leveraging either default or add-on password protection software is protecting sensitive information. Then someone attempts to enter different passwords to unlock the device in a brute-force attack. In order to accelerate the effort, he utilizes a computer to emulate keyboard input, vastly accelerating the capability to enumerate possible passwords. However, the keyboard emulation software must recognize when an incorrect password is used and it must initiate the sending of the next password. The addition of speakers, often including a port for headphones, helps address this problem as illustrated in the following discussion. Because most PDAs generate a warning sound when an incorrect password is entered, a short program can be written that enables the cracking computer, once connected to the PDA audio port, to analyze sounds to manage the brute-force attack. The program operates by repeating the following steps until a password is found:

1. The program generates the first (or next) password to try.
2. It keys in the password characters and submits the password.
3. It listens for the sound output. If one is not heard, the password is correct. If a sound is heard, the GUI "wrong password" warning is closed.
4. Retry at step one, which generates the next password to try.

This method allows a substantially accelerated brute-force attack against a PDA password. The effectiveness of this attack, as with many others, depends on the strength of the password. As a general rule, keyboards enable up to 30 characters per second. Thus, if a password consists of eight characters maximum, a keyboard emulation still only enables the enumeration of approximately four passwords per second. If the password is 4 characters long using alphabetic characters only (52 symbols), it will be cracked in 11 days ($524 \times 4/30 = \sim 1,000,000$ seconds, or ~ 11 days). This model is effective depending on the expected useful life of the information being sought. Many devices such as the PocketPC enable the use of a four-digit *personal ID number* (PIN); the time to crack these devices is far less than the previous example. The number of possible PINs is 10,000 (numeric only), and thus the time to enumerate is 20 minutes ($10,000 \times {}^4\!/_{30} = \sim1,333$ seconds$= \sim20$ minutes). This scenario highlights the obvious importance of password policies.

Infrared Ports: Understanding the Variety of Threats   Most PDAs come standard with highly integrated wireless ports based on an infrared "beam." These ports have become the most common wireless technology built into PDAs, offering convenient short-range connectivity and data exchange. However, these ports expose an effective means of compromising even strong third-party security solutions on PDA devices and need to be well understood from a threat perspective.

IrDA ports, as they are referred to, are often used for PDA-to-PDA exchanges of information as well as PDA-to-desktop or server synchronization. IrDA v1.0 enables data exchanges at speeds up to 115,200 *bits per second* (bps), and the new standard IrDA v2.0 has a maximum transfer rate of 4 million bps. These speeds approach those of LANs and transfer large amounts of information in a short amount of time. Using IrDA eliminates the need to have different connection cables between different kinds of PDAs, which other wireless technologies do as well, offering a valuable extensibility within organizations.

However, this comes with a cost; the following outlines some of the key risks of allowing wide use of IrDA from a security point of view:

- IrDA traffic may be intercepted.
- Some PDAs enable transmitted code to auto-execute upon receipt via IrDA (Trojan).
- Some PDAs enable data to be retrieved by another device via IrDA.

IrDA Traffic May Be Intercepted   IrDA uses infrared that may be received by another IR sensor. Thus, traffic between two devices may be recorded and analyzed. This means an IrDA channel cannot be used to transfer private data without encryption or another means of ensuring privacy. If you intend to use an IrDA channel for a secure data transfer, a secure network protocol should be considered.

Fortunately, the IrDA channel has a limited range, as anyone who has used a standard television remote control has probably observed; devices communicating with IrDA must be placed close to each other. Often, for adequate transfer rates, the maximum distance between the devices is 1 meter, and in some cases the distance should be less than 10 centimeters (4 inches). The good news is, at these distances, you will most likely notice if someone suspicious is trying to intercept your data. With proper training, a user can virtually eliminate the risk that a device will be compromised by the use of the IrDA port. However, this does not address the risk that the IrDA port introduces for compromising a device if it is lost or stolen.

Auto-Executing Transferred Code (Trojan Horse)   Although most PDAs enable a wide array of information to be exchanged via the IrDA port, the Palm OS provides an instructive example by enabling a user to easily exchange a complete application between two devices. The key security risk is based on the default handling after a device has received the application via IrDA. After receiving an application, the Palm OS executes this application in order to complete its initialization. At this point, the application has administrative access to the PDA, so it also has the capability to run malicious code. For example, a Trojan application may collect and transfer user information back to the original PDA, retrieve a user password, or disable applications on the device.

Retrieval of Data from Another PDA   Previous Palm OS versions (v3.5.2 and earlier) have weaknesses that allow another device to retrieve and recover user passwords. These older Palm OS systems use a relatively simple algorithm to convert the user password into a scrambled form for system storage and future authentication purposes. The weakness of the algorithm are that it is two-way, allowing for easy conversion of the scrambled password back to the original password once the algorithm is known. One algorithm is used for passwords four characters or less in length; another algorithm exists for passwords greater than four characters. The scrambled password is calculated from the original password and a 32-byte hard-coded block using an *exclusive OR* (XOR) operation, as described in Chapter 7, "*Understanding Handheld Security Is Largely About the Operating System.*" As shown later in this section, a PalmCrypt utility, by @stake (www.atstake.com), implements both algorithms for encryption and decryption of the password. See the @stake report for full details about Palm OS password encryption algorithms.

Palm OS devices use a special application called HotSync for data synchronization between the PDA and the desktop. HotSync is able to use different channels for data transfers:

- Serial COM port (via a cradle).
- USB port (via a cradle).
- IrDA channel.
- Network connection.

The following example demonstrates how to retrieve and recover a device's user password via the IrDA channel. First, we need to set up HotSync to use the IrDA port instead of the serial port on the device. Then we will launch the HotSync application and choose the IR connection to the PC computer, as illustrated in Figure 8-1. On the desktop, enable the IrDA for the desktop HotSync application, as illustrated in Figure 8-2. After setup, you will be able to synchronize the Palm device with the desktop via the IrDA channel.

The @stake group published a detailed security report on how to retrieve a scrambled password from another device and recover the original user password from the scrambled form. This report can be found at www.atstake.com/research/advisories/2000/a092600-1.txt.

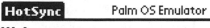

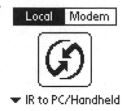

FIGURE 8-1. Palm setup of IrDA.

FIGURE 8-2. Desktop setup for IrDA.

This weakness of the Palm OS password storage enables the HotSync application to transmit the scrambled password during its connection with the desktop via any channel. It also further enables the data block to be intercepted and decoded. The data block has the structure illustrated in Table 8-1.

This block, received by the desktop as part of the HotSync operation, enables HotSync to prompt the user for his or her password. To fully illustrate this exploit, let's examine how to use a utility program to capture and recover a password from one device by using another. @stake uncovered this possibility and published a utility that enabled this scenario called *NotSync*.

To demonstrate how to retrieve a user password from a Palm device, two Palm devices are required: one with Palm OS 3.5.2 or an earlier version (the target device) and another device with any Palm OS version greater than Palm OS v3.0 (the interception device). NotSync should be installed on the interception device, as illustrated in Figure 8-3, and the target device should be configured to use HotSync via IrDA, as illustrated in Figure 8-4. Place the two devices head to head, and start NotSync

TABLE 8-1   Data Block Structure

DATA LENGTH (BYTES)	DATA TYPE
4	Header
6	Execution buffer
4	User ID
4	Viewer ID
4	LastSyncPC ID
8	Successful synchronization date
8	Last synchronization date
1	Username length
1	Password length
128	Username
128	User password

NotSync

Waiting to connect with target

FIGURE 8-3.   Running NotSync application.

HotSync

Connecting with the desktop using IR to PC/Handheld

(Cancel)

FIGURE 8-4.   Running IrDA HotSync.

on the interception device. Start HotSync on the target device, and press the graphical HotSync button on the target device. After establishing a connection, NotSync receives the data block containing the user password from the target device, and decodes and displays the user password.

This technique requires physical access to an unlocked target Palm device to initiate a HotSync session manually, and the target device must be running Palm OS version 3.5.2 or earlier. However, the same technique could be used to recover a Palm user password over a Network HotSync, in the Enterprise LAN, under similar circumstances. In this case, a network sniffer could be used to recover the scrambled password.

Devices running newer versions of the Palm OS store the password in an MD5 hash, instead of a weak scrambled password model. We strongly recommend that you use the most recent versions of devices and OSs in general. Specifically, with the Palm OS, we encourage you to seek Palm OS 4.0 or higher. (It is possible to upgrade older devices to Palm OS 4.1 because Palm, Inc. sells upgrade packs with Flash ROM to Palm OS 4.1, and almost all previous Palm models are upgradeable.)

PDA and Wireless Networks: Cellular, 802.11x, and Bluetooth

Enterprise users are increasingly using PDA devices and terminals connected to the Enterprise LANs via wireless networks. Thus, wireless network security becomes part of the PDA security challenge. As devices continue to improve with an increased combination of *central processing unit* (CPU), memory, and now bandwidth, more important applications are becoming commonplace on Enterprise PDAs, introducing a whole new class of threat. Devices are no longer primarily a risk due to the information they store, manage, and transmit. Increasingly, they are becoming a point of access to a wide variety of resources that exist within the organization's private network. IT departments face a nearly daily expansion of the nature and number of wireless-enabled devices configured to access resources within trusted networks.

PDAs are now using 802.11x/*Wireless Fidelity* (WiFi), *Code Division Multiple Access* (CDMA), *Global System for Mobile Communications* (GSM)/*General Packet Radio Service* (GPRS), Bluetooth, and other *wireless LANs* (WLANs), often including multiple networking technologies within a single, mobile PDA device. These PDAs can leverage public and private networks from a wide variety of physical locations due to their wireless capabilities, limiting the effectiveness of physical security measures. The primary wireless networks in use are based on radio transmissions for data versus the use of cables for wired networks.

All the technologies we're discussing receive and transmit information using electromagnetic waves. Wireless technologies use wavelengths ranging from the *radio frequency* (RF) band up through the IR band. The frequencies in the RF band cover a significant portion of the electromagnetic waves' radiation spectrum, as illustrated in Figure 8-5, extending from 9 *kilohertz* (kHz), the lowest allocated wireless frequency, to 1,000's of *gigahertz* (GHz). The network technologies in use cover a wide range on the RF spectrum and Table 8-2 illustrates the usage of electromagnetic spectrum today.

Categories of Wireless Networks   Wireless networks are frequently categorized into three groups based on the coverage range:

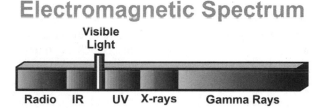

FIGURE 8-5.   Electromagnetic waves' radiation spectrum.

TABLE 8-2.   Electromagnetic Spectrum Usage

FREQUENCY RANGE	MILITARY USES	COMPETING USES
138–144 MHz	Land mobile radio, Tactical air/ground/air	Little *low earth orbit satellites* (LEOS), public safety
225–400 MHz	Tactical air/ground/air, data links, satellite command, military air traffic control, search and rescue, executive command, tactical command	LEOS, public safety, terrestrial digital audio broadcast, CMRS
400.15–401 MHz	Defense meteorological satellite program	Mobile satellite service

TABLE 8-2.   Electromagnetic Spectrum Usage *(Continued)*

FREQUENCY RANGE	MILITARY USES	COMPETING USES
420–450 MHz	Ballistic missile surveillance and early warning radars, shipboard/airborne early warning radars, missile/air vehicle flight termination, air vehicle command links, troop position location, anti-stealth radar, foliage penetration radar	Auxiliary broadcast CMRS, biomedical telemetry wireless local loop
1215–1390 MHz	Long/medium range air defense, radio navigation, air route surveillance radars, tactical communications, test range support, air/fleet defense, drug interdiction, *global positioning system* (GPS), remote satellite sensors, nuclear detection	Mobile satellite service, GPS, general wireless communications systems, wind profile radars
1435–1525 MHz	Telemetry supporting, entire air industry	Digital audio broadcast: satellite and terrestrial, mobile satellite service
1755–1850 MHz	*Department of Defense* (DoD) satellite tracking, telemetry and command, point-to-point microwave, air combat training systems, tactical command, tactical data links	Personal communication systems, multipoint distribution systems
2200–2290 MHz	DoD satellite tracking, telemetry and command, guided missile telemetry, point-to-point microwave	Personal communication systems
3100–3650 MHz	High-power mobile radars, shipboard air traffic control, missile links, airborne station keeping	Multipoint distribution systems, wireless local loop, fixed satellite service
4400–4990 MHz	Fixed wideband command, mobile wideband command, command links, data links	General wireless communications systems, fixed satellite service, public safety

- *WWAN* are wireless wide area networks and include wide coverage area technologies such as *second generation* (2G), *third-generation* (3G) cellular, *Cellular Digital Packet Data* (CDPD), GSM, Mobitex, and DataTAC.
- *WLAN* are wireless local area networks, including 802.11x, Hyperlan, and several others.
- *WPAN* are wireless personal area networks such as Bluetooth and IR.

Wireless Wide Area Networks   WWAN has become largely driven by cellular providers as well as pager networks. The emerging uses tend to be Internet access over cellular networks for web-based applications and always-on email driven by the lower-bandwidth but highly available pager networks. Moving forward with phone-convergent devices and increasing bandwidth, many emerging applications are expected to substantially expand the role of WWAN platforms.

Wireless Local Area Networks   WLANs enable a greater flexibility and portability than traditional wired LANs. Unlike the traditional LAN, which requires a wire to connect the device to the network, the WLAN connects computers to the network using access point devices. Access point devices provide a wired access point for wireless client devices equipped with wireless network adapters. 802.11x-type access point devices typically have a coverage range of up to 100 meters. This coverage area is called a cell or range. Users move freely within cells using their notebook or PDA device. Access point cells can be linked together to allow users an even broader "roam" capability.

Wireless PAN or Ad Hoc Networks   *Ad hoc* networks, such as Bluetooth, are networks that dynamically connect remote devices such as cell phones, notebooks, and PDAs, as illustrated in Figure 8-6. These networks are called *ad hoc* because of their sniffing capabilities. In contrast to fixed WLANs' network infrastructure, *ad hoc* networks maintain dynamic network configurations, relying on a system of mobile routers connected by

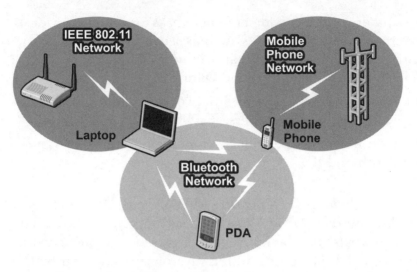

FIGURE 8-6.   Bluetooth integration with other networks.

wireless links, enabling the devices to communicate. As devices move about in an unpredictable fashion, these networks must be reconfigured on-the-fly to handle topography changes. The mobile router is integrated in a device like a PDA; this router ensures that a mobile device, such as phone, stays connected to the network. The router maintains the connection and controls the flow of communication.

Wireless Standards and the Security They Implement Many competing wireless standards have implemented standards for addressing security in varying degrees. *Advanced Mobile Phone System standard* (AMPS) is a first-generation mobile phone standard and enables devices from different manufacturers to communicate with each other. This standard uses *Frequency Division Multiple Access* (FDMA). Other telephony standards include the *Time Division Multiple Access* (TDMA) standard, CDMA standard, and GSM. Above these standards, many of the cell phones and wireless-enabled PDAs support the *Wireless Application Protocol* (WAP), which provides secure access to email and other Internet resources.

Cellular Wireless Networks and GSM Security   At this time, many cellular WAN networks are primarily used for voice transmission purposes, better known as phone calls, yet cell networks can also be used for data transmission purposes. However, only the latest standards, such as CDMA 2000 and GSM/GPRS, offer support for high-speed data transfers and security implementations, including authentication and data flow encryption.

GSM/GPRS Security Approach Overview   The security objective for the GSM system is to ensure the security of the public switched telephone networks. The use of radio frequencies, as the transmission media, exposes a number of threats. The GSM *Memorandum of Understanding* (MoU) Group produces guidance for operator interaction with its members. The technical features for security are only a small part of the security requirements; the greatest threats come from simpler attacks, such as disclosure of the encryption keys, insecure billing systems, or data corruption. Existing cellular systems have a number of potential weaknesses that were considered in the security requirements for GSM. The security for GSM must be appropriate for the system operator and customer as follows:

- Operators must ensure that they issue bills to the right people, and that the services are not compromised.
- Customer privacy is established, preventing traffic from being intercepted and overheard.

The countermeasures are designed as follows:

- To make the radio path as secure as the fixed network, which implies anonymity and confidentiality.
- To have strong authentication, and to protect the operator against billing fraud.
- To prevent operators from compromising each other's security, because of competitive pressures.

The security must not:

- Significantly add to the delay of the initial call setup for subsequent communication.
- Increase the bandwidth of the channel.
- Allow for increasing error rates.
- Add excessive complexity to the rest of the system.
- Add expensive overhead (it must be cost-effective).

The design of an operator GSM system must take into account the environment and have secure procedures such as:

- The generation and distribution of keys.
- The exchange of information between operators.
- The confidentiality of the algorithms.

GSM/GPRS Security Approach Technical Review   The following sections discuss the security services provided by GSM:

- Anonymity
- Authentication
- Signaling protection
- User data protection

ANONYMITY   Using a temporary ID provides anonymity. When a user switches on his or her GSM device, the real ID is used, and a temporary ID is issued. Only by tracking the user is it possible to determine the current temporary ID.

AUTHENTICATION   Authentication is used to identify the user, or the holder of the smart card, to the network operator. Authentication is performed by a challenge and response mechanism. A random challenge is issued to the module; the mobile device encrypts the challenge using GSM's authentication algorithm (A3) and the key assigned to the mobile device. The mobile device then sends the response back to the operator, which checks that, given the key of the module, the response to the challenge is correct. Transmitted data doesn't contain any useful information because a new random challenge will be used the next time.

Authentication is performed via the following process as illustrated in the process map in Figure 8-7. A 128-bit random number is generated by the network and sent to the mobile device. This number, or R, equals a plain text random number. The mobile device uses this random number R as the input for creating a 32-bit encrypted response. To create the encrypted response, a secret key, or Ki, will be used. The 32-bit response or *signed response* (SRES) is the cipher text and is sent back to the network.

User Data and Signaling Protection   The response is passed through the GSM's A8 algorithm both on the mobile device and on the network to derive the key Kc, which is used for encryption of the signaling and messages to provide privacy with the A5 series crypto-algorithm. Random value R and Ki are combined via A8 algorithm to produce the 64-bit encryption key for the A5 cipher; 114-bit data frames are encrypted using the key and frame number using the A5 algorithm.

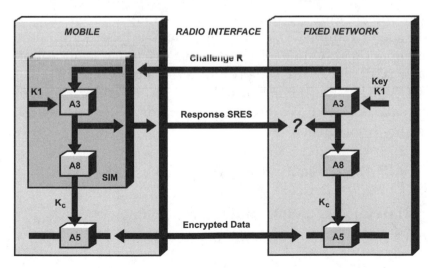

FIGURE 8-7   Authentication process.

IMPLEMENTATION AND ROAMING   The A3 authentication algorithm is the operator's option and is implemented within the smart card, known as a Subscriber Interface Module (SIM) card. In order for the operator to work without revealing the authentication algorithm and mobile keys (Ki) to each other, GSM enables triplets of challenges (R), responses (SRES), and communication keys (Ki) to be sent among operators over the connecting networks.

The A5 series algorithms are contained within the mobile equipment, and they have to be sufficiently fast. Two defined algorithms are used in GSM known as A5/1 and A5/2.

The A3 and A5 algorithms are combined, usually as COMP-128. The enhanced Phase 1 specification, developed by the *European Telecommunications Standards Institute* (ETSI), enables internetworking between mobile devices containing A5/1, A5/2, and unencrypted networks. These algorithms can all be built using a few thousand transistors and usually take up a small area of the chip within the mobile device.

WORLDWIDE USE OF THESE ALGORITHMS   GSM offers three different data security options: unencrypted transmission, the A5/1 algorithm, and the A5/2 algorithm. These options arose because the GSM standard was designed for Western Europe, and export regulations did not allow the use of the original technology outside Europe. The GSM MoU Group controls the use of the algorithm in the network operator's infrastructure, and members of Conference of European Post and Telecommunications (CEPT) can use the present A5/1 algorithm.

COMP-128 was compromised in 1998, the collective security weaknesses that represent the compromise consist of the following components:

- COMP-128 is weak as it allows extracting IMSI and Ki keys from the device:
  - Direct access to SIM card (phone cloning).
  - Over-the-air queries to the phone.

- COMP-128 algorithm has an internal weakness that allows the cipher text to be cracked.
- COMP-128 key implementation was weakened as 10 bits of the generated keys are set to 0.

Mobitex and DataTAC   Mobitex and DataTAC are WWAN networks and are used by AT&T, Cingular, and Motorola. The data transfer has a maximum rate of 10 Kbps. No data encryption takes place, and thus wireless devices should implement data encryption at a higher level. The RIM 800 and 900 series by *Research In Motion* (RIM) are examples of devices that use the Mobitex and DataTAC networks, providing encryption on the high-level protocols. Another disadvantage of these networks is that they don't support *Transmission Control Protocol/Internet Protocol* (TCP/IP) protocols directly.

802.11x Understanding the Range of Security Issues 802.11x WLANs offer a nonwired communication method that makes them very attractive. They offer organizations and home users increased efficiency and cost savings. The benefits can include the following:

- *User mobility*   No physical connection to the network is required. The mobile user can have a high-speed network connection.
- *Rapid installation*   The installation time is reduced due to the absence of wires.
- *Flexibility*   A quick install is performed and existing WLANs are removed.
- *Scalability*   Network topologies can easily be reconfigured and scaled from small peer-to-peer networks to large networks, enabling the roaming over a broad area.

History and Overview of 802.11x WLAN's Architecture   The *Institute of Electrical and Electronics Engineers* (IEEE) initiated the 802.11 project under the terms "to develop a *Medium Access Control* (MAC) and *Physical layer* (PHY) specification for wireless connectivity for fixed, portable, and moving stations within an area." The goal was to create a standards-based technology that could span multiple physical encoding types and applications, similar to what was done with the 802.3 Ethernet standard. The 802.11a uses *Orthogonal Frequency Division Multiplexing* (OFDM) to reduce interference. This *Wireless Fidelity* (Wi-Fi) technology uses the 5 GHz frequency spectrum and has a maximum data transfer rate of 54 Mbps.

IEEE developed the 802.11b standards to provide wireless technology, such as wired Ethernet, that has been available for many years. The 802.11b WLAN operates in the unlicensed 2.4 to 2.5 GHz frequency band, and 802.11a WLAN operates in the 5 GHz frequency band. The standard 802.11 characteristics are shown in Table 8-3.

TABLE 8-3.   Standard 802.11 Characteristics

CHARACTERISTIC	DESCRIPTION
Physical layer	*Direct Sequence Spread Spectrum* (DSSS), *Frequency Hopping Spread Spectrum* (FHSS), and IR.
Frequency band	2.4 GHz (ISM band) and 5 GHz.
Data rates	1 Mbps, 2 Mbps, 5.5 Mbps, 11 Mbps (11b), 54 Mbps (11a), 54 Mbps (11g).
Data and network security	RC4-based stream encryption algorithm for confidentiality, authentication, and integrity. Limited key management.
Operating range	About 150 feet indoors and 1,500 feet outdoors.
Throughput	Up to 11 Mbps (54 Mbps planned).
Positive aspects	Ethernet speeds can be utilized without wires; many different products from many different companies are available. Wireless client cards and access point costs are decreasing.
Negative aspects	Native mode has poor security; throughput decreases with distance and load.

Two types of equipment are used in the WLAN: wireless stations and access points. The wireless station is a client device, usually a notebook or PDA with a wireless *network interface card* (NIC). The wireless network card can be a card that is inserted into the card slot (*Personal Computer Memory Card International Association* [PCMCIA], Compact Flash, Secure Digital, or Sony Memory Stick), or an integrated network card. Many manufactured notebooks and PDAs are equipped with an integrated wireless network card. The NICs use a radio or an infrared beam to establish the connection as a bridge from a WLAN to a wired LAN, as illustrated in Figure 8-8, or in a point-to-point configuration, as shown in Figure 8-9. Access points are the base station for the WLAN, which enables wireless clients to communicate with each other and obtain access to resources outside of the WLAN.

Access points also provide a bridging function. Bridging connects several networks together, allowing them to communicate. Bridging uses either point-to-point or multipoint configurations. With point-to-point architecture, the two LANs are

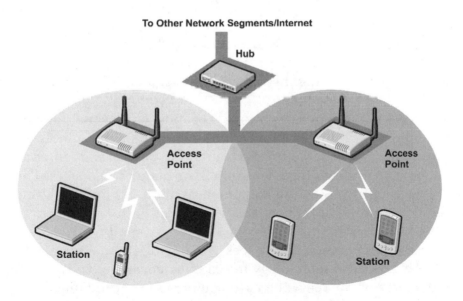

FIGURE 8-8. Example topology of an integrated wireless and wired network.

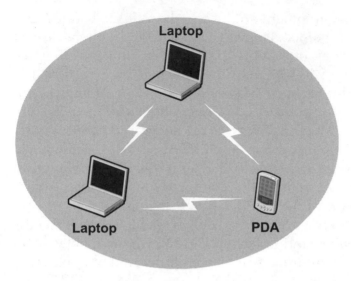

FIGURE 8-9.   Example topology of a point-to-point wireless network.

connected to each other via the LAN's respective access points. Multipoint bridging means that one subnet on a LAN is connected to several other subnets, each via an access point. For example, if subnets A, B, C, and D are used, subnet A's access point would connect to the B, C, and D access points.

Bridging may be used to connect LANs between different buildings or campuses. Bridging access points are typically placed on top of buildings where the distance is several miles between the access points.

The WLAN coverage range depends on several factors, including the transfer data rate, the source of radio interference, physical characteristics, and the power and antenna usage. Theoretical ranges are from 29 to about 500 meters. The typical range for 802.11b equipment is about 50 meters. Using omnidirectional antennae, this range can be increased up to 400 meters with illustrations in Figure 8-10 and 8-11 characterizing these connection spaces. A range of 400 meters makes WLAN the ideal technology for campus coverage, and special antennas may increase the coverage range to several miles.

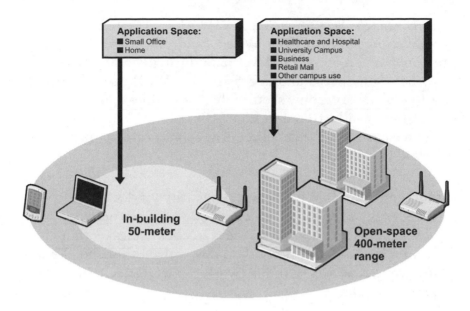

FIGURE 8-10. Between building distance examples.

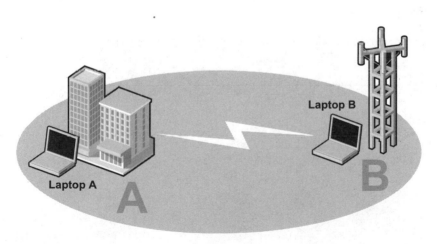

FIGURE 8-11. Building to open space distance examples.

WEP and the Security of 802.11x WLANs   The IEEE 802.11b specification identifies the *Wired Equivalent Privacy* (WEP) protocol that protects link-level data during wireless transmissions between clients and access points. WEP doesn't provide end-to-end security, only security for the wireless portion of the connection, as illustrated in Figure 8-12.

Three basic WEP-related security services are defined by the standard for WLANs:

- *Authentication*   WEP provides security services to verify the identity of the wireless clients. It provides network access control by denying access to client stations that cannot authenticate properly. Thus, only authorized clients are granted access to the network.

- *Confidentiality or privacy*   WEP provides privacy matching that of a wired network. The goal is to prevent information from being compromised by passive attacks. WEP allows data access to authorized clients only.

- *Integrity*   WEP is intended to ensure that messages are not modified during transmission between wireless clients and the access point in an active attack.

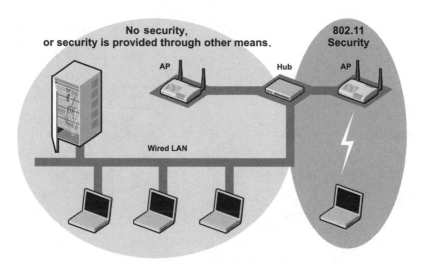

FIGURE 8-12.   WEP security role.

The standard does not address other security services, such as audits, authorization, and nonrepudiation.

AUTHENTICATION   The standard describes two ways of validating wireless users attempting to get access to a WLAN as illustrated in Figure 8.13. The first method is cryptography based. For the other method, identifying a wireless client attempting to join a network can be done in two different ways, both of which are referred to as identified-based verification mechanism. In the first identified-based method, the wireless client requesting responds with a *Service Set Identifier* (SSID) of the wireless network, alternatively the access point can be set to not require an SSID—in either case there is not real authentication. This range of methods in aggregate are referred to as *open system authentication,* as illustrated in Figure 8-13.

With open-system authentication, a client is validated if it simply responds with an empty string for the SSID, or NULL authentication. With the second method, close authentication, the wireless device must respond with an actual SSID of the wireless network. Access is granted if the client responds with the correct string (from 0 to 32 bytes) identifying the *Basic Ser-*

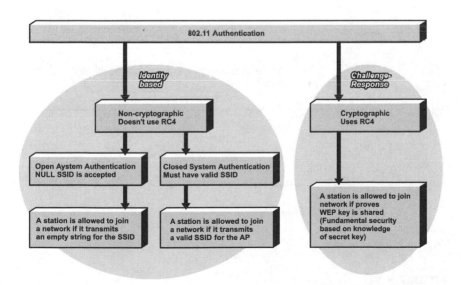

FIGURE 8-13.   Open-system authentication.

vice Set (BSS) of the wireless network. Practically, neither of these two schemes offer robust security against unauthorized access. Open and close authentication are highly vulnerable to attacks.

Shared-key authentication is the cryptographic technique for authentication. It is a challenge-response scheme based on checking if the client has knowledge of a shared secret. A random challenge is generated by the access point and sent to the wireless client. The client, using the cryptographic key (WEP key) shared with the access point, encrypts the challenge and returns the result to the access point. The access point decrypts the result computed by the client and allows access only if the decrypted values are the same as the random original challenge. Similar authentication is used in the GSM/GPRS networks, as illustrated in Figure 8-14.

This method uses a RC4 stream cryptoalgorithm. Note that the authentication method just described is a rudimentary cryp-

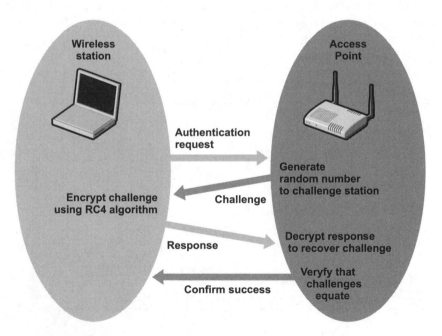

FIGURE 8-14.   Challenge response.

tographic technique and doesn't provide mutual authentication. The client doesn't authenticate the access point, so no guarantee exists that the client is communicating with a legitimate access point and a wireless network. Also, it is a well-know fact that unilateral challenge-response schemes are known to be weak. They fall victim to several different forms of attack, including the "man-in-the-middle" attack where the attacker intercepts packets prior to arriving at the intended destination and substitutes alternate packets.

PRIVACY   The 802.11b standard supports privacy by using cryptographic techniques. The WEP uses RC4 symmetric keys and a stream cipher to generate a random data sequence. This key stream is added via XOR to the transmitted data. Through the WEP technique, data can be protected from disclosure during data transmission over the wireless network. WEP is used to secure all data above the 802.11b WLAN layers, protecting traffic such as TCP/IP, *Internetwork Packet Exchange* (IPX), *Hypertext Transfer Protocol* (HTTP), and others.

WEP supports cryptographic key sizes from 40 to 104 bits. In general, increasing the key size increases the security of cryptography. In practice, most WLAN rely on 40-bit keys. However, it is known that WEP privacy is vulnerable to attacks regardless of key size.

INTEGRITY   The 802.11b specification provides data integrity for messages transmitted among wireless clients and access points. The integrity services should reject any message changed by an active adversary "in the middle." This technique uses the *Cyclic Redundancy Check* (CRC). According to Figure 8-15, a CRC-32 or frame check sequence is calculated on each payload prior to transmission. Next, the packet with CRC is encrypted using a RC4 key stream to produce a ciphered message. After receiving the packet, decryption is performed and the CRC is recalculated. If the CRC is not equal to the original one, the message is treated as corrupt. This indicates an integrity violation and the client discards the packet. Unfortunately, the 802.11b integrity is vulnerable to some attacks regardless of key size.

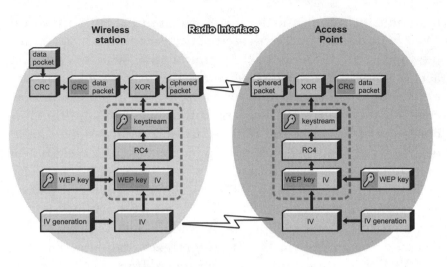

FIGURE 8-15   Client-to-access-point interaction.

The key issue of the 802.11b standard is that the specifications don't offer methods for key management. Thus, the WLAN environment has a lot of vulnerabilities. These vulnerabilities include WEP keys that are nonunique, never changing, factory defaults, or weak keys. For example, keys might be constructed from all zeros or based on short, simple passwords. Also, poor key management for 802.11b inhibits the scaling of WEP-secure WLANs. For example, a large campus may have thousands of clients and access points, and the task of generating, distributing, loading, and managing keys for this environment is extremely demanding.

Problems with IEEE 802.11b Standard Security   This section describes known vulnerabilities in the standard security of 802.11x. For 802.11x, WEP is used for security and uses an RC4 cryptoalgorithm with a variable-length key to protect network traffic. The 802.11b standard supports 40-bits of WEP cryptographic keys. Some vendors implement products with 104 bits of key support, plus an additional 24 bits, resulting in a maximum length of 128 bits. However, keys are often based on user passwords, which may reduce the effective key size.

   Specialists have discovered security issues with the protocol. Issues include passive attacks to decrypt network traffic via statistical analysis, active attacks to insert new packets into the WLAN traffic from unauthorized mobile clients, active attacks to decrypt traffic, and dictionary-building attacks. The dictionary-building attack is possible by analyzing network traffic for a period of time.

   WEP has several problems, however:

- Many users in the wireless network may share an identical key for long time periods of time. Poor key management support in WEP creates this problem. If the PDA or notebook with the WEP key is stolen, all computers in the WLAN could be compromised, because they all use the same WEP key. Also, because every station uses the same WEP key, a large amount of encrypted traffic may be readily available for analytic attack.

- The *initialization vector* (IV) is a 24-bit field sent in the clear text portion of a message, as illustrated in Figure 8-15. This 24-bit string, used to initialize the key stream generated by the RC4 algorithm, is a relatively small field used for cryptography. Reusing the same initialization vector produces identical key streams for data protection, and the short IV means that the key stream will repeat after a relatively short time on a busy network. The 802.11b standard doesn't specify how initialization vectors are set or changed, and wireless cards from the same vendor may generate the same IV sequence or even generate a constant IV. As a result, hackers can record and analyze the network traffic, recover the key stream, and decrypt the ciphered traffic.

- The initialization vector is part of the RC4 encryption key. Thus, hackers can see 24 bits of every packet key, and when combined with knowledge about the weaknesses in the RC4 key, this information stream simplifies an analytic attack. The end result is that the hacker can recover the encryption keys after analyzing a relatively small amount of traffic.

- WEP doesn't provide cryptographic integrity protection. Also, the 802.11 MAC protocol uses a noncryptographic CRC to check the integrity of packets and acknowledges packets with the correct check sum. This combination of a noncryptographic checksum with stream ciphers is potentially dangerous and often leads to unintended "side channel" attacks in which "other" information is looked at in order to break a key or other security. Also, an active attack permits the attacker to decrypt any packet by systematically modifying the packet; the CRC sends it to the wireless access point and notes whether the packet has been acknowledged. These attacks are crafty; some experts say it is risky to use encryption protocols without cryptography integrity protection, because the possibility of interactions with other protocol levels allows access to the cipher text.

Table 8-4 summarizes security issues of the 802.11b standard protocol which are plagued by a range of design problems that go beyond just a weakness in the cryptographic algorithm.

Security Requirements and Threats   This section covers in general the risks to security, attacks on the confidentiality, and integrity and network availability, as illustrated in Figure 8-16.

Security attacks are typically divided into passive and active attacks:

- *Passive attack*   This allows an unauthorized client to simply gain access to the network, but who doesn't modify any data (eavesdropping). The two top passive attacks are
  - *Eavesdropping*   The attacker simply records network transmissions between client(s) and access point(s).
  - *Traffic analysis*   The attacker uses more sophisticated methods to monitor and analyze data transmissions.

TABLE 8-4.   802.11b Standard Protocol Security Issues

SECURITY ISSUE	COMMENTS
Security features are frequently not enabled.	Security features in the vendor products are not enabled sometimes when shipped and the user may forget to enable them.
Initialization vectors are short or static.	A 24-bit initialization vector can cause a generated key stream to repeat. It makes cryptoanalysis easier.
Cryptographic keys are short.	Forty-bit keys are not long enough. The keys should be greater than 80 bits in length. The longer key lessens the chance of a successful brute-force attack on the key.
Cryptographic keys are shared.	Keys that are shared can compromise a system.
Cryptographic keys cannot be updated automatically and frequently.	Cryptographic keys should be changed often to prevent brute-force attacks.
RC4 has a weak schedule, as is inappropriately used in WEP.	An efficient attack may be performed using the "visible" 24-bit IV and the weakness of an initial few bytes of the RC4 key stream.
Packet integrity is poor.	CRC32 is not reliable for providing crypto-graphic integrity, and message modification is possible. Cryptographic protection is required to prevent attacks. Use of noncryp-tographic protocols facilitates attacks against cryptography.
No user authentication.	Only the device is authenticated. A stolen device may be used for network access.
Authentication is not enabled, only simple SSID identification occurs.	Identity-based systems are highly vulnerable, particularly in a wireless system.
Device authentication is a simple shared-key challenge-response.	One-way challenge-response authentication is subject to man-in-the-middle attacks. Mutual authentication is required.

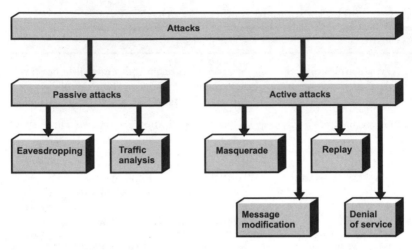

FIGURE 8-16.   Attack types.

- *Active attack*   Unauthorized clients make modifications to a message, data stream, and so on. An active attack can be detected but not prevented. The four forms of active attacks are as follows:
  - *Masquerading*   The attacker imitates an authorized user.
  - *Replay*   The attacker monitors transmissions and retransmits the messages as an authorized user.
  - *Message modification*   The attacker modifies legitimate messages by deleting, adding, changing, or reordering them.
  - *Denial of service*   The attacker prevents the normal use of management communications facilities, many times by flooding the network with millions of packets.

All security risks are a result of exploits enabled by the 802.11x protocol. The results of the attacks may include a loss of confidential information, legal and recovery costs, and a loss of network service.

Loss of Confidentiality   Confidentiality guarantees that information is not made public or disclosed to unauthorized persons. This is a fundamental security requirement for most organizations. Due to the nature of radio transmissions in the WLAN environment, a significant effort is required to ensure confidentiality. Wire cable protects physical media to ensure no unauthorized access occurs, whereas radio transmissions provide a different challenge. This makes traditional security techniques less effective.

Passive eavesdropping on the wireless network may cause a significant risk to the organization. Hackers can listen and obtain sensitive information, including proprietary information such as network IDs, passwords, and configuration data. This risk is caused by the fact that radio transmissions may be available outside a building perimeter, or an insider may intercept network traffic. This risk is increasing due to new installations of 802.11x access points in parking lots, roads, and even in cars. An attack may be easy because users may not have WLAN security at all or because the 802.11b WLAN vulnerabilities compromise the system.

A network analyzer tool, known as a sniffer, performs these types of attacks. Public tools, such as AirSnort and WEPcrack, are also available online for automatic network traffic analysis. AirSnort takes advantage of weaknesses found in the key-scheduling algorithm of RC4, which is part of the WEP standard. Only a Linux-based computer with a wireless card is required. AirSnort passively monitors WLAN packets and computes the encryption key. Approximately 100MB of network traffic are required to get enough information to recover the key. The time needed to complete the recovery ranges from several hours to several days, depending on the network load level. Busy Wi-Fi networks may take less than an hour. After collecting the network traffic, the encryption key is calculated in just a few seconds. After the encryption key is recovered, all packets on the LAN are wide open.

Wireless access points are generally connected to the wired LANs. If organizations use hubs, someone may connect an unauthorized notebook into the same hub as the access point. Then the notebook (with the NIC in promiscuous mode) may monitor all traffic intended for wireless clients in the WLAN, because the hub broadcasts all network traffic to all connected devices. To avoid this risk, organizations should use switches instead of hubs.

Network traffic-monitoring software (sniffers) may obtain user names, passwords, and other important authorization information when they are sent over wireless connections. Intruders may masquerade as a legitimate user and obtain access to the wired network from an access point. After connecting to the wired network, an intruder can scan the network using publicly available sniffing software. This software grabs usernames, passwords, IP address maps, and anything else needed to obtain access to data.

The risk of a rogue access point also exists. An insider may place a rogue access point into a hidden place within the building and obtain access to the network. If this access point is configured with a stronger signal than the existing access point, this rogue access point can intercept the wireless traffic between an authorized access point and a wireless client.

LOSS OF INTEGRITY   Integrity issues within the wireless networks are similar to those faced by wired networks. It is difficult to achieve integrity without cryptographic data protection. Unauthorized data manipulations are possible within organizations. Because the 802.11b standard doesn't provide strong message integrity, it enables attacks that may compromise system integrity. Due to simple CRC integrity checks, attacks focused on simple message modification are possible threats.

LOSS OF NETWORK AVAILABILITY   A malicious user may use radiation of a powerful signal to overwhelm a legitimate wireless signal, known as *jamming*. Authorized users can inadvertently cause a *denial of service* (DoS) attack using this technique. For example, transmitting a large amount of data for a long time may cause an access point to be inaccessible to other users.

OTHER SECURITY RISKS   Actually, connecting to the Enterprise network using publicly available WLANs can be done in many ways. Airports and some coffee franchises, such as Starbucks, provide wireless access to Internet, which may be used to connect to an Enterprise network. These outside networks introduce new risks, however:

- Public networks are accessible to anyone, including hackers.
- They are the bridges for other networks, and they potentially allow others to perform attacks on other networks.
- Use of a high-power radio transmission level allows for easier eavesdropping.

Using a connection to Enterprise networks via an untrusted public network creates vulnerabilities for an Enterprise unless it takes proper security precautions. Organizations should protect resources with an application-layer security protocol such as *Transport Layer Security* (TLS) or the *Internet Engineering Task Force* (IETF) version of *Secure Socket Layer* (SSL).

Risk Mitigation With the increasing introduction of PDAs depending on wireless access points to integrate with a corporate network, an organization must determine what level of risk is acceptable. Security requires money for security equipment, maintenance, and operating expenses.

Technical Countermeasures Technical countermeasures involving software include: properly configured access points, updating software on a regular basis, authentication solutions, performing security audits, and effective encryption. These methods will ensure that PDAs are properly protected, and the configuration information embedded on them (which supports authenticated access to the network) will be protected in the event of a lost or stolen device.

Access Point Configuration Access points should be configured using a concrete security policy. To avoid weak defaults in vendor factory settings, the administrator should configure an administrator password, encryption settings, a reset function, an automatic connection function, an Ethernet MAC, an *access control list* (ACL), shared keys, and *Simple Network Management Protocol* (SNMP) agents.

Updating the Default Password Each WLAN device has its own factory (default) settings that compose security vulnerabilities because they are well known. The administrator password is a key example. On some access point devices, the administrator password is empty, and an unauthorized user can easily obtain access to an access point if no password exists. Administrators need to change default settings according to the active security policy, which must have strong administrative password requirements. If security is strict, organizations should utilize a password generator.

As an alternative to password authentication, two-factor authentication can be implemented. One form of two-factor authentication uses a symmetric key encryption algorithm to generate a new code every minute. This code is aired with the user's personal code, or PIN, for authentication. Another example of two-factor authorization is pairing a user's smart card and PIN. This requires a hardware smart card reader or an authentication server for the PIN, and certain commercial products are available to implement this capability. However, using an automated password generator or two-factor authentication depends on the organization's security requirements and budget constraints.

Set Proper Encryption Level   Encryption settings should be set to the strongest level available that is appropriate for the organizational security requirements. Typically, access points have few encryption settings available. Here are the possible settings:

- None.
- 40-bit shared key.
- 128-bit shared key.
- WEP encryption uses a simple stream cipher generation and XOR operation, which doesn't require many computing resources.

Organizations should choose the longest available encryption key. Although WEP has weaknesses, as described previously, those weaknesses are not dependent on the key length.

Reset Function   The reset function is a security issue. It allows a person to reset the administrator security settings for an access point. Manually resetting the access point returns all settings to the factory default, including security settings, and all administrator settings will be lost. The default settings usually don't require an administrator password and may disable encryption. If a malicious user gets physical access to the device, he may exploit the reset features and drop the security settings of the access point. The administrator should prevent physical access to the access points to mitigate this threat. Security audits also help to solve this security issue.

Using MAC ACL Functionality   The NIC MAC address uniquely identifies each computer on the network. Networks use MAC addresses to regulate communications between different computer NICs on the network subnet. Usually, 802.11 products provide capabilities to restrict access to the WLAN using MAC ACLs that are stored and distributed across many access points. The MAC ACL grants or denies access to the computer, using settings for the specified MAC addresses in the list.

However, MAC ACL doesn't represent a strong mechanism. MAC addresses are transmitted in a clear-text form, and they may be captured with ease. Malicious users can spoof MAC addresses, changing them on their own computer so that the MAC address is valid for the WLAN. MAC ACL may give an additional level of security but should be used with caution. Although it may be effective enough against passive eavesdropping, the process of establishing and maintaining MAC ACL in a medium or large network is relatively expensive.

Changing the SSID   Access points may have a default SSID value. The actual SSID may be captured in the WLAN, but the default SSID value adds vulnerability to the system.

Changing Default Cryptographic Keys   An access point may have preset encryption keys designated by the manufacturer. The default encryption keys can be a serious security flaw, because these values are well known and are the same in a wide range of access point models. Choose an appropriate encryption key as follows:

- Use the longest available key length, such as 128-bit keys, if the system allows them.
- Change encryption keys often. Static encryption keys may be recovered and the security may be compromised.

Changing the Default SNMP Parameter   SNMP agents have access points that allow the access points and clients on the WLAN to be monitored. Usually, the default value of the SNMP community string is public, and assigned privileges are read only, or read and write. Using default values for the community string adds vulnerability to the system because the string is well known. If a malicious user gets an access-to-access SNMP point, he or she will be able to write data to the access point. If SNMP is required, the organization should change the SNMP community string often, with privileges set to read only. If SNMP is not required, the SNMP agents should be disabled.

DHCP Issue   The Dynamic Host Control Protocol (DHCP) provides an automatic connection to the network. The DHCP server assigns IP addresses automatically to devices associated with an access point. DHCP assigns addresses to workstations as needed from an established IP address range. The server assigns a dynamic IP address to the device as long as encryption settings are compatible with WLAN settings. The main threat is that any user may connect a PDA to the WLAN with DHCP, and the valid IP address will be assigned to the new device because the DHCP server may not know which wireless devices require access to WLAN. Risk mitigation requires the disabling of DHCP and a string of statically assigned IP addresses instead of dynamic IP addresses.

Disabling DHCP for large Enterprise networks may be costly, however, due to great administration efforts required to manage static IP addresses on larger networks. One possible solution is using a DHCP server inside a wired LAN firewall, which grants access to the wireless network located outside the wired LAN firewall. Another solution is the use of access points with integrated firewalls.

Software Patches and Updates   Vendors always try to fix bugs and eliminate vulnerabilities in initial software and hardware releases, as well as in follow-up updates and patches. Network administrators need to check regularly to see if a vendor has published security patches or upgrades for existing software. A *National Institute of Standards and Technology* (NIST) ICAT[1] vulnerability database at http://icat.nist.gov lists all known security exposures in current software and hardware products.

For example, RSA Security, Inc., developed a fix for security holes in WEP. This enhancement, referred to as *fast packet keying*, generates a unique key to encrypt each packet on the WLAN. The IEEE approved it as a fix for the 802.11 protocol, and vendors have developed patches for many existing products, as shown on the NIST web site.

Personal Firewalls   As we have discussed, resources on a wireless network have a higher risk of attack than on a wired LAN. Personal firewalls offer additional protection against such attacks on a wireless network; these firewalls are client-managed or centrally managed software-based solutions. A client-managed version is intended for low-end users because each user configures the firewall without a central security policy. Centrally managed solutions provide better protection because the security policy is applied for all users. Some of the firewalls have *virtual private network* (VPN) and audit capabilities. Personal firewalls don't protect against all attacks but provide an additional level of protection. Although these products are just emerging in the PDA area, companies like Bluefire are providing integrated solutions that we should continue to expect from broader security providers over time.

---

[1]ICAT is a searchable index of information on computer vulnerabilities. It provides seach capability at a fine granularity and links users to vulnerability and patch information. The ICAT Metabase is a product of the Computer Security Division at the National Institute of Standards and Technology.

Intrusion Detection System   An *intrusion detection system*, or IDS, is an effective tool against unauthorized attempts to gain access to the WLAN. An IDS for WLAN may be host- or network-based.

A host-based IDS is available in limited solutions for the PDA, such as the Bluefire software mentioned earlier. Some IDSs may stop an attack on the system, but the main functions of an IDS are system monitoring and alerts.

Network-based IDS solutions may inter-operate well with PDA communities to provide monitoring, auditing, and the enforcement of access rules, as the offerings from Sygate do. Their general role is to monitor LAN network traffic in real time (as much as possible) and compare the actual traffic with pre-defined attack patterns. For example, certain attacks send highly fragmented packets over the network to target a system using known system vulnerabilities. Some systems may crash after receiving too many fragmented packets (such as Windows 95 or 98). Voidozor is an example of an attack application. An Network-based IDS recognizes packets that conform to the attack pattern and take appropriate actions, such as alerting the administrator and dropping the attacker's network connection.

A host-based IDS has an advantage over a network-based one because it can monitor network traffic before encryption and take appropriate high-level action. A network-based IDS monitors encrypted traffic (SSL or VPN) but cannot decrypt the transmitted data.

Encryption   802.11 wireless access points generally have only three encryption modes:

· None
· 40-bit key
· 104-bit key

The "none" mode represents a high risk because a network transmits clean non-encrypted data that can be sniffed and altered. The 40-bit shared key encrypts network traffic but is not strong enough, because in a brute-force attack a powerful computer can break the 40-bit encryption. The 104-bit key encryption is more secure, but unfortunately WEP has weaknesses that make the encryption vulnerable regardless of the key length used. Any organization should use the maximum available key length and check regularly for encryption-related software patches and updates. However, on a PDA, CPU limitations may need to be factored into whether a 104-bit key is practical for users on the network.

Security Assessments   Security assessments or audits are required to check the security level of the system on a regular basis. It is important to perform regular audits of the WLAN using network analyzers and other such tools. Use network analyzers to ensure that access points are transmitting properly using correct channels. The administrator should also regularly check the WLAN area for rogue access points and other unauthorized access. Third-party companies may be used for security checks due to their specializations.

Hardware Solutions   The hardware for mitigating WLAN risks includes smart cards, VPNs, *public key infrastructure* (PKI), biometrics, and other hardware solutions.

Emerging Security Standards and Technologies   The IETF and IEEE are working on improving the 802.11 standard and WLAN security. Several initiatives have been developed.

The first proposal represents a significant modification to the existing IEEE 802.11 standard. The *IEEE 802.11 Task Group* (TGi) is defining a second version of WEP that uses a new encryption standard, *Advanced Encryption Standard* (AES). When available, the AES-based solution will provide a robust solution, but it will also require new hardware and protocol changes.

The second initiative is improving WEP in the short term. TGi is currently defining the *Temporal Key Integrity Protocol* (TKIP) that will enable software upgrades, by themselves, to solve the WEP issue (no hardware changes are required). The main goal of TKIP is to remove all known WEP vulnerabilities and to enable its operation on existing Wi-Fi hardware.

The third initiative is the introduction of the new standard, IEEE 802.1x. This standard allows the access point to authenticate a particular NIC by consulting an authentication server. The 802.1x standard supports authentication servers such as the *Remote Authentication Dial-In User Service* (RADIUS) or Kerberos. The 802.1x standard can be implemented with different *Extensible Authentication Protocol* (EAP) types, including EAP-MD5 for Ethernet LANs and EAP *Transport-Level Security* (TLS) for 802.11b WLANs. The 802.1x standard addresses WEP vulnerabilities by securely delivering session keys. It makes keys obtained from one WEP session useless for another WEP session.

Wireless LAN Security Checklist   Table 8-5 presents a list of recommendations for creating and maintaining the 802.11 WLAN.

TABLE 8-5.   802.11 WLAN Recommendations

	RECOMMENDATION	IMPORTANCE
1.	Develop a security policy to address 802.11 wireless technology.	Must have
2.	WLAN users should be well trained in computer security awareness and the risks associated with wireless technology.	Must have
3.	NIC and access point firmware must be upgraded with the latest software patches and updates to ensure that security patches are applied immediately.	Must have
4.	Perform security checks and audits on a regular basis.	Must have
5.	Ensure external boundary protection around the perimeter of the organization's building(s).	Must have
6.	Deploy physical access controls to the building and other security areas using photo IDs, smart card readers, and so on.	Must have
7.	Take a complete inventory of all access points and 802.11 wireless devices.	Must have
8.	Perform tests of the actual WLAN coverage area.	Must have
9.	Ensure access point channels are at least five channels different from nearby wireless networks to avoid interference.	Must have

TABLE 8-5. 802.11 WLAN Recommendations (*Continued*)

	RECOMMENDATION	IMPORTANCE
10.	Locate access points inside buildings that are not near exterior walls and windows.	Must have
11.	Place access points in secured areas to avoid u292 authorized physical access to the hardware.	Must have
12.	Turn off access points when they are not in use.	Must have
13.	Make sure only authorized persons perform access point resets.	Must have
14.	Apply security policy on the access point after a reset.	Must have
15.	Change the default SSID on the access point	Must have
16.	Disable the broadcast SSID feature on the access points.	Must have
17.	Validate that the SSID string is not equal to the organization name.	Must have
18.	Disable the broadcast beacon of the access points.	Good to have
19.	Ensure that default access point settings are changed.	Must have
20.	Disable all insecure and nonrequired protocols on the access points.	Must have
21.	Enable all security features on the access points and wireless clients, including cryptographic authentication and WEP.	Must have
22.	Use 128-bit keys if possible.	Must have
23.	Default encryption keys must be replaced by unique keys.	Must have
24.	Install a properly configured firewall between wired LAN and WLAN.	Must have
25.	Install antivirus software on wireless clients.	Good to have
26.	Install personal firewalls on all wireless clients.	Good to have
	RECOMMENDATION	IMPORTANCE
27.	Deploy a MAC ACL.	Good to have
28.	Use network switches instead of hubs for access point connection.	Good to have

TABLE 8-5. 802.11 WLAN Recommendations (*Continued*)

	RECOMMENDATION	IMPORTANCE
29.	Deploy VPN technology based on *IP Security* (IPSec) for wireless communications.	Good to have
30.	Test and deploy software patches and upgrades on a regular basis.	Must have
31.	Make sure all access points have strong administrator passwords.	Must have
32.	Make sure all passwords are changed regularly.	Must have
33.	Deploy user authentication such as biometrics, smart cards, two-factor authentication, or PKI.	Good to have
34.	Ensure that 802.11 *ad hoc* network mode is disabled.	Must have
35.	Use static IP addresses on the WLAN.	Good to have
36.	Disable DHCP.	Good to have
37.	Enable user authentication for access point management.	Must have
38.	Ensure that management traffic utilizes only the wired LAN for access.	Good to have
39.	Use robust community strings for SNMP agents.	Must have
40.	Configure SNMP settings on access points for the minimum privilege level: read only.	Must have
41.	Disable SNMP if it is not used.	Must have
42.	Use SNMPv3 or its equivalent to provide better security of management traffic.	Good to have
43.	Use a built-in serial port for access point management.	Good to have
44.	Use enhanced authentication for a WLAN such as RADIUS and Kerberos.	Good to have
45.	Deploy intrusion detection sensors on the WLAN to detect attacks.	Good to have
46.	Deploy an 802.11 security product that offers enhanced cryptography or authentication.	Good to have
47.	Fully understand the impacts of deploying any security feature or product prior to deployment.	Must have
48.	Track 802.11 security products, standards (IEEE or IETF), threats, and vulnerabilities.	Good to have

Wireless LAN Risks and Security   In addition to reviewing the broad security check list for establishing a wireless network as illustrated in Table 8-5, Table 8-6 highlights some commonly encountered security threats and introduces sample approaches to remediate them.

TABLE 8-6.   Wireless LAN Risks and Security

SECURITY SUBJECT	SECURITY THREAT	POSSIBLE SOLUTION
Access point cells	Eavesdropping on a WLAN may be performed either inside or outside the WLAN building.	Implement encryption to reduce the risk of eavesdropping and control the coverage range of the WLAN.
Access point building-to-building bridging	Eavesdropping on a WLAN can be performed outside of WLAN building.	Implement encryption to reduce the risk of WLAN eavesdropping.
Encryption settings	Access point default settings usually don't enable encryption.	Enable 128-bit key encryption, if possible.
Reset function	A reset may cause the actual security settings to be dropped and the default vulnerable settings to be restored.	Return access point settings to the appropriate values.
Shared key authentication	Many vendors use the same default shared key, which could allow an unauthorized user to gain access to the network.	Change encryption keys to unique values and choose WLAN products with better key management.
Access point administrative password	An access point without an administrator password allows access by an unauthorized user. Implement a strong password policy (at least eight characters in length with alphanumeric characters).	Always protect access points with a password. Restrict remote administration from a particular computer or subnet of the network.
Ethernet MAC ACL	A MAC address is transferred in clear form without encryption.	Understand limitations of MAC ACLs.

TABLE 8-6. Wireless LAN Risks and Security (*Continued*)

SECURITY SUBJECT	SECURITY THREAT	POSSIBLE SOLUTION
Encryption	Some algorithms use longer keys; some use shorter keys.	Implement the highest level of encryption. WEP must be enabled, but due to WEP vulnerabilities, use of the VPN is also highly recommended.
Security patches and updates	The current version of the software or firmware may contain bugs or weaknesses that may be used by a hacker.	Apply the latest patches or updates from the vendor on a regular basis.

Bluetooth   Bluetooth is a network technology for wireless communications. The *Bluetooth Special Interest Group* (BSIG) developed it in 1998, and BSIG is composed of more than 1,500 participating companies, including Ericsson, Nokia, Intel, IBM, Toshiba, Microsoft, 3Com, and Motorola. Bluetooth can be used to connect one device to another (such as a PDA and a cell phone) without cables for use in small *ad hoc* networks called *piconets*. Bluetooth technology should make our lives easier because it simplifies connections between various devices: desktops, notebooks, PDAs, mobile phones, printers, and so on. Just by pressing several buttons on the screen, we have wireless networks in action. However, if it is so easy to connect one device to another, wouldn't that make data easy to steal? The answer is that the Bluetooth standard was designed to establish secure connections between devices. Let's examine the Bluetooth standard in more detail.

The Bluetooth protocol architecture consists of several high-level protocols, as illustrated in Figure 8-17, including WAP, TCP/IP, the *user datagram protocol* (UDP), and the *Point-to-Point Protocol* (PPP). It supports cable replacement protocol RFCOMM, which provides emulation of serial ports over the L2CAP protocol, and telephony adapter protocols. These protocols are not Bluetooth-specific but were adopted for use in the Bluetooth architecture. The main advantage of such architec-

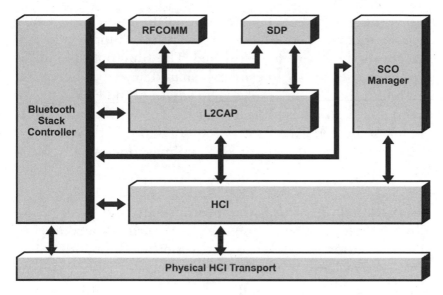

FIGURE 8-17.   Bluetooth architecture.

ture is that Bluetooth can integrate with existing applications and transport protocols.

Here are a few brief descriptions of some protocols:

- *Secure Discovery Protocol (SDP)*   This enables a Bluetooth device to discover and gather information about other Bluetooth devices (device type, services, and so on).
- *Logical Link Control and Adaptation Protocol (L2CAP)* This adapts upper-level protocols over the *Host Controller Interface* (HCI) level.
- *HCI*   This is a low-level Bluetooth controller interface.
- *RFCOMM*   This is a cable replacement interface.

The Bluetooth specifications support security features at the link level. It supports device authentication and data encryption, and it is based on a shared secret between a pair of devices. A pairing procedure is used to generate a shared secret when devices establish communication for the first time. To keep transmissions from breaking up, Bluetooth employs frequency-hopping, the practice of skipping around the radio band 1,600 times each second. Each Bluetooth device has a unique address, which allows the device to be tracked. The Bluetooth initialization process uses a PIN to establish shared keys for encrypting Bluetooth traffic among devices. The length of the PIN can vary between 1 and 16 bytes. A four-digit PIN code is usually sufficient, but some applications may require a longer PIN.

Security Modes   Bluetooth devices may operate on three security modes:

- *Security Mode 1*   This is the most insecure mode; the Bluetooth device doesn't initiate any security procedure. Bluetooth uses this mode to discover other Bluetooth devices in order to initiate a connection.
- *Security Mode 2*   This mode enforces security after establishing a link between devices at the L2CAP level. This mode enables flexible security policies involving the application layer and the controls running in parallel with the lower protocols.
- *Security Mode 3*   This is link-level-enforced security. The difference between Security Mode 2 and Security Mode 3 is that the security mode initiates a security procedure before the channel is established.

Key Management   The Bluetooth standard describes how devices should generate encryption keys to encrypt transmitted data. During the initialization process, the device generates a set of keys, and the device's PIN is an important part of the negotiation process.

Encryption   The Bluetooth device encryption manager encrypts a payload, or the content, of the packets. The encryption key size varies from 8 to 128 bits, and devices negotiate the proper key size. Each device has a parameter, which determines the maximum key size. One device sends a suggested key size to another device, which accepts or sends back its own key size proposal. This process continues until a consensus is reached or a device aborts the negotiation. If the master encryption key is used, three encryption modes are possible:

- Nothing is encrypted.
- Only individually addressed traffic is encrypted using a master key; broadcast traffic is not encrypted.
- All traffic is encrypted using a master key.

Security Issues   The Bluetooth encryption scheme has some weaknesses. For instance, the use of PIN codes is annoying because they must be entered each time a connection is established. If you have a network of Bluetooth devices, this could be troublesome. However, the strength of the initialization keys is equal to the strength of the PIN codes. A 4-digit PIN has 10,000 possible values. Manufacturers set the default PIN to 0000, and if the user forgets to change the PIN, it becomes a security hole. The device manufacturer defines the encryption level, so administrators should review the implementation prior to concluding how secure a particular Bluetooth network may be.

Also, the Bluetooth standard doesn't provide for built-in user authentication and authorization; it allows for authentication of the device only. It supports the use of upper-level protocols for service-level user authentication. That is, a Bluetooth connection can be established without real authentication, but accessing a specific service can only be done with user authentication.

Note that unlike the IrDA standard, the Bluetooth standard allows data to be retrieved from the target device without user interaction under some circumstances. It is a possible to connect a set of trusted devices without user interaction, and the Bluetooth pairing is established automatically. A strong security policy eliminates this issue.

Loss and Theft

PDA devices, cell phones, notebooks, and other handheld computers are subject to loss and theft. Employees may leave their mobile devices out of the organization's building or secure areas by simply forgetting about them. Of course, mobile devices may also be stolen. Certain criminal groups specialize in the theft of mobile devices, such as notebooks and cell phones, for resale.

According to The Insurance Agency, Inc., "Of the 2.3 million insurance claims for notebook computer losses in 2001, 25 percent were theft related; notebook theft has increased by 53 percent since 2000." Organizations should enforce security policies to minimize the loss of data and hardware. Mobile device loss or theft can be prevented in two different ways:

· Mobility restriction.
· Software and hardware tracking.

Mobility restriction means that the mobile device is somehow physically attached to a certain area. For example, a PDA may be tied to the desk, via a steel cable, such as a Kensington lock. The user can also leave a laptop for a while and be sure that it cannot be stolen. However, this doesn't prevent unauthorized access to the laptop, so authentication and data protection tools remain important. However, using mobility restrictions for a PDA defeats the entire purpose of having a PDA.

Tracking software helps to find a lost or stolen device, and a couple of companies offer this solution, such as Absolute Software (*www.absolute.com*). The main idea is that a laptop can be connected to the Internet provider, and the tracking software will identify the laptop location and report it to the tracking station. We expect a similar solution for the PDA to arrive in the near future, as the "connected" PDA (with built-in wireless LAN and phone functionality) becomes more commonplace.

Assessing the Strength of Data Locks   To prevent the loss of information, the PDA should have a locking mechanism via software or hardware tools, or both. The strength of the data locks should be assessed to evaluate the level of protection for the following reasons:

1. To illustrate the range of options for addressing an authentication only (no data encryption) solution, the following illustrates general levels of data lock protection starting with the weakest example and progressing to the strongest solutions.

- No protection. The device doesn't have data locks at all, and unauthorized access is available.
- A simple authentication method using a PIN.
- A password authentication. Any password is allowed.
- A strong alphanumeric password.
- Smart card authentication.
- Biometric authentication.

2. Beyond authentication, encryption can further add to the quality of the data lock, again allowing for a range of options as exemplified below starting from weakest to strongest:

   - Use of a fast proprietary encryption algorithm or XOR method with a key stored on the device.
   - Use of well-known symmetric algorithms such as *Triple DES* (3DES), Blowfish, AES, and others.
   - Using public key encryption algorithms (RSA, DHA, and so on).

Penetrating a Locked Device Once a device is locked, the user cannot access the device resources (applications and data). The device displays only a lock screen with the authentication prompt to enter a PIN or password, or to insert a smart card. If the locked device is stolen and the intruder wants to access the data on the device, he or she must penetrate this protection level. Several possibilities exist:

- Enter a correct key to the authentication system.
- Bypass the authentication system using a different technique.
- Get the data directly from the device using low-level methods, including physically disassembling the device.

Breaking the device protection depends on the qualification of the intruder and the available resources. Perfect protection does not exist, and all protection schemes can theoretically be broken at some point. The only exception is if the data is encrypted via a disposable codebook, and the codebook is destroyed. Infinite time and resources would be required to break data protected in such a manner.

Insider Threats

It is a well-known fact that employees are the most common threat to an organization. Employees have some privileges within the Enterprise network, so if a displeased employee wants to cause damage, his or her goal is easier than an outsider lacking those privileges.

Users Attempting to Bypass Security   Security at any level is inconvenient to the end-user attempting to perform his or her job. Password prompts quickly become annoying, and the user may attempt to turn off security completely. A malicious user or employee may then attempt to gain access to resources beyond his or her rights. This means that the security administrator must be able to deploy a security policy over all organizations, and users should not be able to switch off or modify security settings, which would compromise the company's security policy.

Use of Access to Penetrate the Enterprise   Each user has a set of assigned rights and privileges as assigned by the administrator. Like hackers who attempt Enterprise network access from outside the network, authorized users may attempt similar access inside the Enterprise network. A malicious user may use access to authorized network components in order to access other, initially unreachable Enterprise network resources. All attempts to gain access to the Enterprise network resources should be detected and recorded regardless of their source, either inside or outside the Enterprise network.

POCKETPC-SPECIFIC TOOLS, TIPS, AND TECHNIQUES

Methods for penetrating Windows CE-based devices have been thoroughly investigated. Using weaknesses in the authorization mechanisms or using backdoors to pass through system authorization are two possible methods. We will show you how to bypass security, via ActiveSync and a backdoor, in the PocketPC bestseller: the Compaq (now Hewlett-Packard) iPAQ.

Understanding the Threat from ActiveSync

Windows CE-based devices use Microsoft ActiveSync to synchronize data between the handheld and the desktop. Synchronized software and data include Microsoft Outlook (a *personal information manager* [PIM]) and third-party software, such as AvantGo. ActiveSync enables the use of many synchronization transports. PocketPC devices may use a serial (COM), USB, Infrared (IrDA) port, or a network connection via a LAN (wired Ethernet) or WLAN (Wi-Fi, GSM, CDMA, and others).

To synchronize data between the PocketPC and the desktop, both sides are equipped with ActiveSync software that manages the exchange of data. Unfortunately, Microsoft doesn't publish any protocol specification. Using different tools, such as PortMon by Sysinternal.com or a network sniffer, it is possible to eavesdrop on the ActiveSync process. If the device is locked by a password, ActiveSync prompts the user to enter a correct password to continue synchronization. Otherwise, synchronization does not continue until the correct password is entered.

Also, ActiveSync enables a user to save a password, and ActiveSync does not prompt for a password during the following synchronization. Thus, two vulnerabilities are possible:

- A user password may be intercepted during synchronization.
- A user password may be located on the user's desktop in a scrambled but recoverable form.

Brute-Force Attack   ActiveSync analysis shows that authorization is performed by sending the user password to the device, which then authenticates the password. Next, the device replies that the password is correct, and if it is, it continues with the synchronization process. Otherwise, the device replies that the password is incorrect, and the server may try another password. The number of password attempts is also unrestricted. This means that it is possible to create an application that emulates the ActiveSync server (at least several initial stages of a password check) and then performs a brute-force attack. As we will see, similar applications were created for Palm OS's HotSync.

Such attacks are especially effective against both the older PocketPC devices and the 2002 PocketPC devices where a four-digit PIN is the password. Only 10,000 possible PIN combinations exist, so the attack would not take much time. However, using a strong password with upper- and lowercase alphabetic characters, numeric characters, and special characters with a password length greater than eight letters would make such attacks useless, as discussed at great length in Chapter 7.

Password Sniffing   As stated, the ActiveSync server transmits the user password to the ActiveSync client on the PocketPC device for authentication. ActiveSync does so by using a simple two-way scrambling function. The password is a Unicode string; that is, each character is two bytes and the last character is zero, which is a string terminator.

Let's look at an example of the ActiveSync password scrambling. The device password we'll use will be my key. Converted into a Unicode string in hexadecimal form, it will appear as follows:

$$00\ 6D\ 00\ 79\ 00\ 6B\ 00\ 65\ 00\ 79\ 00\ 00$$

ActiveSync scrambles the password using a bitwise XOR operation with byte 7Bh, which is information stored digitally and is ultimately recorded in 1's and 0's or binary. Let's XOR our password with byte 7Bh, producing:

$$7B\ 16\ 7B\ 02\ 7B\ 10\ 7B\ 1E\ 7B\ 02\ 7B\ 7B$$

This string is transmitted to the device and can be intercepted using a sniffer application. Thus, password recovery is extremely easy. A possible solution would be using third-party synchronization solutions, such as XTND Connect Server and others that use public key authentication and traffic encryption.

Microsoft eMbedded Visual Tools

Microsoft eMbedded Visual Tools enable various Windows CE-based device resources to be monitored, such as:

- System registry
- File system
- Running applications and threads
- Windows objects

Also, MS eMbedded Visual Tools can debug active applications. You can download this software from the Microsoft Web site; visit www.microsoft.com/mobile/developer/ for more information. These tools presuppose access to the device but are invaluable for a range of activities that might be conducted to bypass security on a device and application-level protections or to better evaluate the device-level operating environment. The following provides a brief overview of how to leverage these tools.

Establishing Remote Connection The main difference between the usual desktop utilities and the MS eMbedded Visual Toolkit programs is that each toolkit utility has two parts: a desktop part and a device part. The desktop part provides the *graphical user interface* (GUI), and the device part performs the actual work. Information is transferred between the two parts over a connection between the desktop and the mobile device. Utilities can connect with different device types such as a handheld PC, a PalmSize PC, a PocketPC, a Windows CE.NET device, and others.

Before using the utilities, make sure the connection with the PocketPC device has been properly established; otherwise, the remote utilities will not work. If you run Microsoft eMbedded Visual C++ or Microsoft eMbedded Visual Basic and go to Tools and Configure Platform Manager, the Platform Manager Settings dialog box will be opened, as illustrated in Figure 8-18.

Place the PocketPC device in the cradle and synchronize. When complete, select the appropriate device type from the tree view. For example, if you use an iPAQ 3870, which is a PocketPC-2002-based device, select the PocketPC 2002 (default device) item. Then press the Properties button, and the Device Properties box will open, as illustrated in Figure 8-19.

Check that the Startup Server option is set to Microsoft ActiveSync and press the Test button. The Testing Device Connection dialog box then appears. The Platform Manager establishes the connection with the remote PocketPC device and copies the necessary *Dynamic Link Libraries* (DLLs) to the

FIGURE 8-18.   Platform Manager configuration.

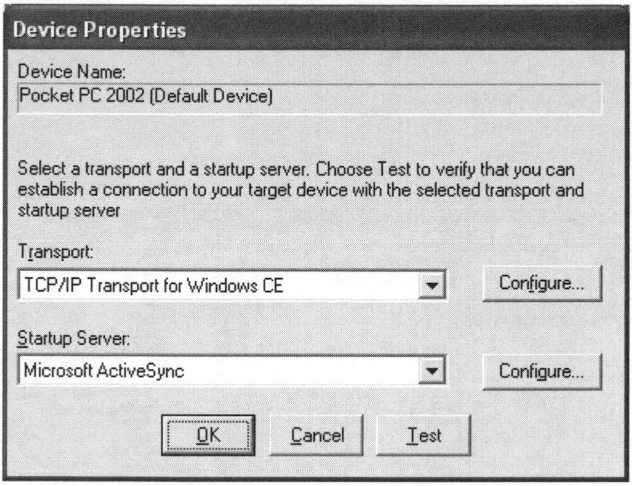

FIGURE 8-19.   Device Properties dialog.

device. If the settings are correct, the message, "Connection to device established," will be displayed.

Inspecting Remote Device Resources   After checking the connection with the remote device, we can use the remote utilities. Unfortunately, the MS eMbedded Visual Toolkit does not create shortcuts for remote utilities. You need to either create shortcuts manually or run them from the *Integrated Development Environment* (IDE) (Visual C++ or Visual Basic) Tools menu. The utilities include:

- Remote Process Viewer enables you to view active process lists, process threads, and loaded module lists, as illustrated in Figure 8-20.
- Remote Spy++ shows windows objects and captures windows message events, as illustrated in Figure 8-21.
- Remote Registry Editor enables you to view and edit the system registry.
- Remote Heap Walker lets you view heap contents for a particular process.

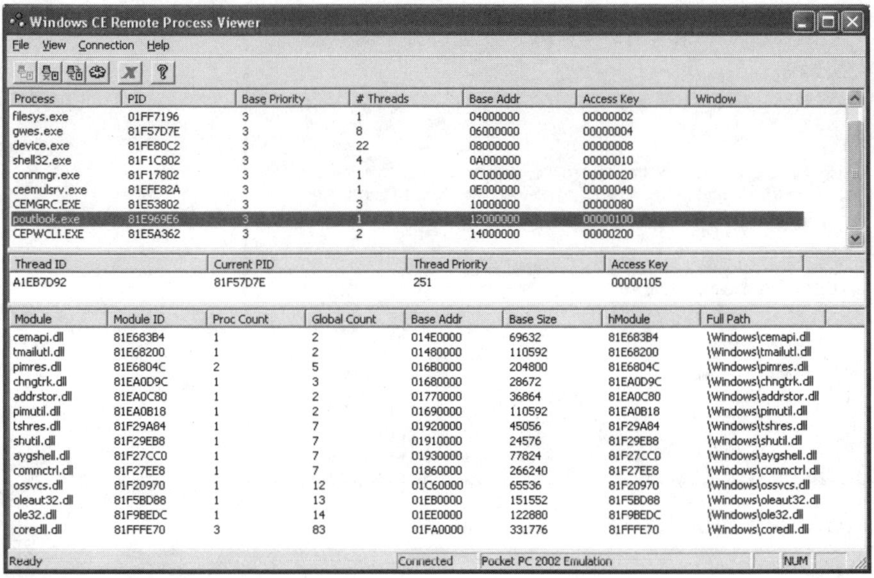

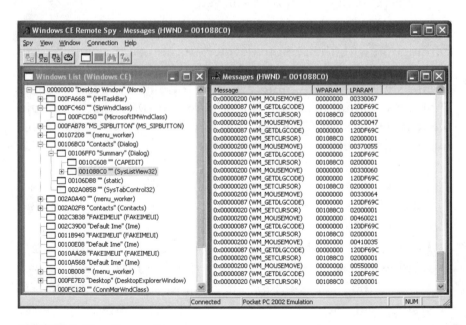

FIGURE 8-20.   Process Viewer.

FIGURE 8-21.   Remote Spy.

- Remote Zoom-in enables you to capture a screen from the remote device.
- Remote File Viewer lets you transfer files with the remote device.

iPAQ Parrot

The Compaq iPAQ is an extremely popular PocketPC device. Some time ago, investigators found that iPAQ firmware has a special utility that can be started before loading the main Windows CE OS. This utility performs almost everything in RAM or ROM, including reflashing, or the process of replacing the information stored on the non-volatile memory with a device or ROM, a totally new OS (such as Linux) onto the ROM. In our case, this tool can be used as the backdoor to access the locked iPAQ. As opposed to Palm OS 4.x or 5.x, which disable debug backdoors on the locked device, the iPAQ built-in tool is a low-level utility that doesn't observe the Windows CE software lock flag. It can be activated on the locked device for access to its data beyond the OS.

The Parrot name comes from one of the oldest devices, the iPAQ 3660, where users found out that if they pressed a certain key combination during a device reset, a picture of a parrot displayed on the screen. At first, this was treated like an "Easter egg," (undocumented functions which are usually intended to be funny) but later people discovered it was more than a picture. When the device is placed in the cradle, which connects to the desktop using the COM port, standard terminal application software (such as HyperTerminal) can communicate with the iPAQ device. Thus, "Parrot" mode is the terminal mode of the iPAQ firmware boot loader.

In this mode, the boot loader can execute certain commands. To see this, an iPAQ with a serial cradle that connects with the COM port on the desktop is required. Modern PocketPC devices, such as iPAQ 37xx, 38xx, 39xx, 19xx, and 54xx models, have a USB-enabled cradle, but it is possible to purchase a cradle with a serial RS-232 connector and modify it to connect to the COM port.

In this example, we will use an iPAQ 3660 with the PocketPC OS modification and a standard serial cradle. To engage, press Reset on your iPAQ while holding the Action button; then release the Action button. Other iPAQ models use slightly different key combinations. Newer versions with a USB interface have additional modes that were intended for ROM upgrades, via USB, without user interaction. If the button sequence is correct, you will see a picture of the parrot on the screen. The latest iPAQ models don't display the parrot picture; they simply display the Compaq logo on the screen. At this time, it is possible to communicate with the iPAQ boot loader.

We can use any OS and terminal application for communication. In this case, we will use Windows to achieve our goal. Check that ActiveSync or any other communication software, which may hold the serial port, is closed, and place iPAQ in the cradle. Start the HyperTerminal and create a new session, as illustrated in Figures 8-22 and 8-23.

Select the appropriate serial port and port settings: 115,200 bps, 8 data bits, no parity, 1 stop bit, and no flow control, as illustrated in Figure 8-24. Press OK and the HyperTerminal

FIGURE 8-22.   Connection Description dialog box.

FIGURE 8-23.   Connect To dialog box 2.

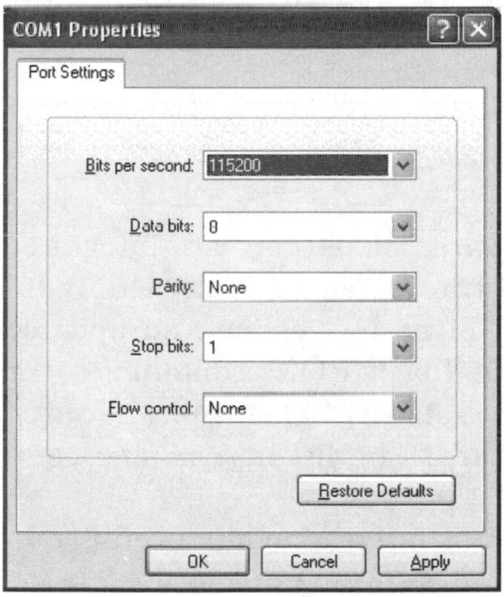

FIGURE 8-24.   COM1 Properties dialog box.

should connect to the iPAQ device. If you don't see anything, press Enter and you should see a prompt like: FW 0:17:25>. Press ? and Enter to see the list of available commands, as illustrated in Figure 8-25.

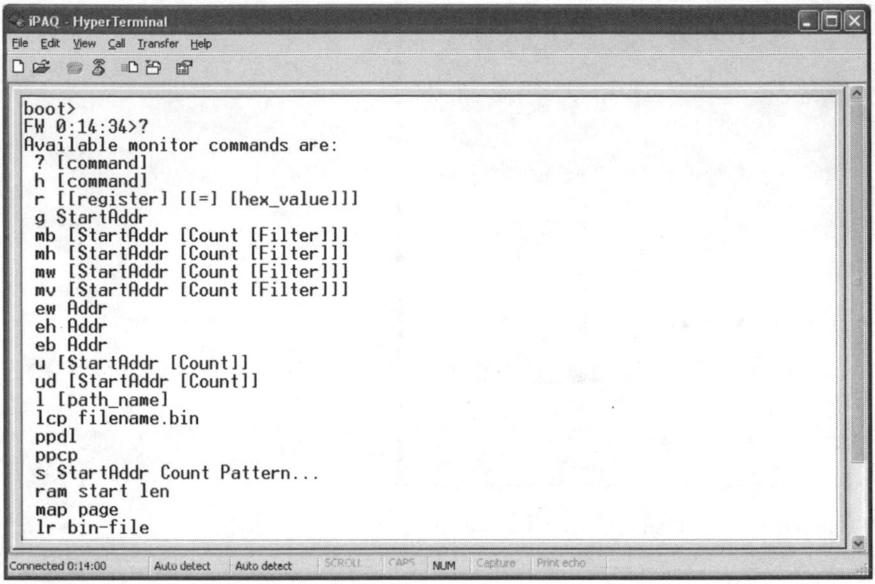
iPAQ - HyperTerminal
File Edit View Call Transfer Help
boot>
FW 0:14:34>?
Available monitor commands are:
? [command]
h [command]
r [[register] [[=] [hex_value]]]
g StartAddr
mb [StartAddr [Count [Filter]]]
mh [StartAddr [Count [Filter]]]
mw [StartAddr [Count [Filter]]]
mv [StartAddr [Count [Filter]]]
ew Addr
eh Addr
eb Addr
u [StartAddr [Count]]
ud [StartAddr [Count]]
l [path_name]
lcp filename.bin
ppdl
ppcp
s StartAddr Count Pattern...
ram start len
map page
lr bin-file
Connected 0:14:00
Auto detect
Auto detect
SCROLL
CAPS
NUM
Capture
Print echo

FIGURE 8-25. Results of the help command in HyperTerminal.

Compaq iPAQ with firmware P1.04 has an integrated boot loader utility called HTC Integrated Re-Flash Utility for Strong ARM (Macaw) v2.31. Newer devices have an updated utility that enables you to save Flash ROM contents to the CompaqFlash card and restore (reflash) ROM from the saved image on the Compaq Flash card. The list of available commands is displayed in Table 8-7.

As we see in Table 8-7, a lot of interesting and useful commands can be used to investigate the iPAQ memory contents. We can read and write at any memory range, seek byte sequences in memory, save Flash memory onto the external memory card, reflash the ROM image, and even load and execute code. Thus, it is possible to retrieve full memory contents from the locked device for further investigation. This is an example of how the authentication system may be bypassed beyond the OS to get access to the device resources.

TABLE 8-7. Commands for Working with the HTC Integrate Re-Flash Utility for Strong ARM (Macaw)

COMMAND	DESCRIPTION
h [command]	Helps on command. When no command is given, it outputs a list of commands.
r [[register] [[=] given, [hex_value]]]	Displays or sets a register value(s). When no register is given, all the registers' content is displayed. When only a register name is given, the content of that register is displayed. If the optional value is also given, the register's content is set to the new value. The= sign is always ignored.
g StartAddr	Jumps and executes from a new address. StartAddr can be either a hex_address or a register name. When StartAddr is not given, the PC is used as the new address. The starting address *must* be in a valid unmapped space. The monitor does not validate this address.
mb [StartAddr either a [Count [Filler]]]	Displays or sets memory content. StartAddr can be hex_address or a registered name. When StartAddr is not given, the memory display continues from the previous address. When the count is not given, a previous count is used for the memory display. The count is initially set to 20 (hex). If the filler is specified, the memory area is filled with the filler value. The memory will be displayed or counted as bytes. StartAddr must be in a valid unmapped space and is not validated.
mh [StartAddr either a [Count [Filler]]]	Displays or sets memory content. StartAddr can be hex_address or a register name. When StartAddr is not given, the memory display continues from the previous address. When the count is not given, a previous count is used for memory display. The count is initially set to 20 (hex). If the filler is specified, the memory area is filled with a filler value. The memory is displayed or counted as half-words. StartAddr must be in a valid unmapped space and is not validated.
mw [StartAddr [Count [Filler]]]	Displays or sets memory content. StartAddr can be either a hex_address or a register name. When StartAddr is not given, the memory display continues from the previous address. When the count is not given, a previous count is used for memory display. The count is

TABLE 8-7.   Commands for Working with the HTC Integrate Re-Flash
Utility for Strong ARM (Macaw) (*Continued*)

COMMAND	DESCRIPTION
	initially set to 20 (hex). If the filler is specified, the memory area is filled with a filler value. The memory is displayed or counted as words. StartAddr must be in a valid unmapped space and is not validated.
mv SourceAddr DestAddr Length	SourceAddr is the hex memory address of the source, DestAddr is the hex memory address of the destination, and Length is the length of the half-word memory data to move.
ew Addr, eh Addr, eb Addr	Addr is the hex memory address.
u [StartAddr [Count]], ud [StartAddr [Count]]	Unassembles instructions. StartAddr can be either a hex_address or a register name. When StartAddr is not given, unassembling continues from the previous address. When the count is not given, a previous count is used. The count is initially set to 14 (hex). For the first unassemble command, EPC is used if StartAddr is not given. StartAddr must be in valid unmapped space and it is not validated. To avoid confusion, all the hex numbers displayed are prefixed with 0x. The absolute target address in a jump or branch instruction is calculated and displayed (except for jr instructions). Offset in offset(base) is displayed in hex format.
l [path_name]	Downloads the BIN file from the bidirectional parallel port. When the path_name is not given, the file to be downloaded is determined by ppfs on the host. Otherwise, path_name on the host is downloaded regardless of the ppfs setting. The file must be in the BIN format (preprocessed SRE). The code is auto-launched once downloaded.
lcp filename.bin	Compares image with Flash by serial port.
lb [path_name]	Downloads BIN file from the bidirectional parallel port. When path_name is not given, the file to be downloaded is determined by ppfs on the host. Otherwise, path_name on the host is downloaded regardless of the ppfs setting. The file must be in the BIN format (preprocessed SRE). Auto-launch is disabled after downloading.
ppdl	Downloads the BIN file assigned by the ppsh command line. This download is done via parallel port.
ppcp	Compares image differences between the download and the Flash data. This command resembles the ppdl command.

TABLE 8-7. Commands for Working with the HTC Integrate Re-Flash Utility for Strong ARM (Macaw) (*Continued*)

COMMAND	DESCRIPTION
s StartAddr a Count Pattern . . .	Searches memory for a pattern. StartAddr can be either hex_address or a register name. The starting address must be in a valid unmapped space. The monitor does not validate this address. Count and StartAddr define a search region. Pattern can be hex numbers or (single or double) quoted strings. A hex number with less than three digits is considered a byte. A hex number with less than five digits but greater than two digits is consider a half-word. Otherwise, a hex number must contain less than nine digits and is considered a word. Up to eight patterns can be given in the command line. They are concatenated as a single search pattern.
ram start len	A DRAM test.
map	Displays a virtual address mapping table.
page	Sets Flash ROM to Page mode.
lr bin-file	Loads BIN to RAM.
cp reg# OPC_2 CRm [value]	Accesses coprocessor registers.
romcopy [RamAddr, ule RomAddr, Length]	Copies the bootloader image of a bootable Flash module to another one without finding the BIN file in the host.
writest [block], between wbuftest [block], eratest [block]	One block equals 256KB. Block input should be 1 and 127.
stress count(Hex)	Writes six kinds of patterns to flash each count for a stress test. Count indicates how many loops you want to run. Count is considered heximal, not decimal.
r2c	Downloads a Windows CE image from Flash to CF.
r2ca	Downloads a Windows CE and a bootloader image from Flash to CF.
c2r	Downloads a Windows CE image from CF to Flash.
erase start len	Erases all Flash memory except the bootloader area.
lcdtest [loop delay(ms)]	The default is for loop to equal 1 and for delay to equal 1000.
romchip [number of chip]	Gets the current number of ROM chip settings if the argument was not specified. It sets the number of ROM chips if the argument was specified.

PALM-SPECIFIC TOOLS, TIPS, AND TECHNIQUES

To improve the security of the Palm OS-based device, we should understand how these devices may be penetrated using different tools. Differing techniques may be used:

- Using developer and debugging tools.
- Resetting the device.

Developer Tools

Developer tools are powerful enough to perform an attack on a locked device. Palm OS Debugger v3.6d7 is a powerful tool that can be obtained from the Palm OS *software development kit* (SDK).

Palm Debugger connects to the Palm device, allowing code to be disassembled, the active application to be debugged, memory contents to be reviewed and changed, and so on (see Figure 8-26). Note that the Palm device uses the Motorola DragonBall-based CPU, so you need some knowledge of the MK68000 assembler before using this tool.

Connecting the Debugger to the Palm Device   The Palm OS responds to a number of hidden shortcuts for debugging programs. To enter a shortcut, follow these steps:

1. On your device, draw the shortcut symbol. This is a lowercase cursive L character, as illustrated in Figure 8-27.
2. Tap the stylus twice to generate a dot.
3. Write the shortcut character, as illustrated in Figure 8-28, using command 3 from Table 8-8.

For example, enter the shortcut illustrated in Figure 8-28 and described in Table 8-8 as character/command 3 in order to disable the automatic power-off feature.

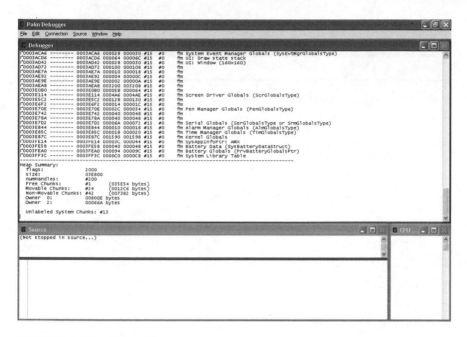

FIGURE 8-26.   Palm debugger.

FIGURE 8-27.   Graffiti shortcut symbol.

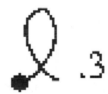

FIGURE 8-28.   Executing command 3 illustrated in Table 8-8.

NOTE

Many of the debugging shortcuts leave the device in a mode that requires a soft reset. To perform a soft reset, press the reset button on the back of the handheld with a blunt instrument, such as a paper clip.

TABLE 8-8.   Example of Debug Commands Available for Use on the Palm OS

CHARACTER	DESCRIPTION	NOTES
1	The device enters debugger mode and waits for the low-level debugger to connect. A flashing square appears in the top-left corner of the device.	This mode opens a serial port that drains power over time. You must perform a soft reset to exit this mode.
2	The device enters console mode and waits for communication, typically from a high-level debugger.	This mode opens a serial port that drains power over time. You must perform a soft reset to exit this mode.
3	The device's automatic power-off feature is disabled.	You can still use the device's power button to power it on and off. You must perform a soft reset to exit this mode.
4	Displays the user's name.	Displays the user entered "User Name" on the device if one has been entered.
5	Erases the user's name and ID.	Warning! When the device is synchronized after using this shortcut, the HotSync Manager application thinks the device has never been synchronized before. This means that records will be duplicated unless you first perform a hard reset (press the reset button while holding the power key).
6	Displays the ROM build date and build time.	Displays the ROM build date.
7	Switches between *nickel-cadmium* (NiCad) and alkaline battery curves to adjust when the battery warnings appear.	This is of limited effectiveness. Low battery warnings do not work well with NiCd batteries.
8	Toggles the backlight mode on a Palm IIIx or Palm V.	One mode is the default back-lighting mode of the Palm IIIx and Palm V. The display is inverted when the backlight is

TABLE 8-8.   Example of Debug Commands Available for Use on the
Palm OS (*Continued*)

CHARACTER	DESCRIPTION	NOTES
		turned on. The other mode is the default backlighting mode of the Palm III where the display is not inverted when the backlight is turned on.
		Restoring a Palm IIIx or Palm V from an existing user may overwrite the shortcut database and cause this shortcut to be lost.
t	Toggles IR test loopback mode.	This mode is useful for debugging applications that beam data. Enabling this mode causes beamed data to be immediately received on the same device (the IR hardware is bypassed). This is useful for testing your beaming or exchange manager code. However, enabling this mode causes more stack space to be used than a normal beam.
i	Initiates a beam receive.	Using this shortcut causes the device to begin waiting for a beam.
s	Toggles IR serial mode.	This shortcut toggles the output port of IR data between the built-in IR port and the serial port.

Place the Palm device in the cradle and write debug short-cut 1. The Palm device debug mode will turn on and wait for the external debugger connection. Run Palm Debugger and choose the appropriate connection type from the Connection menu. Let's look at an example of this using Palm Vx with Palm OS v3.5. In the debugger console window, type the att command. You will see something like the following:

```
EXCEPTION ID = $A0

    10C0D512  *MOVEQ.L   #$01,D0                    | 7001
```

Palm Debugger disassembles instructions in the following order:

```
<address> <instruction mnemonic> <operands> | <instruction
bytes>
```

Breaking into the Locked Palm　Place the locked Palm into the cradle and run the debugger. Write on the Palm debug (shortcut 2 in Table 8-8) to activate the debug console. Make sure Hot-Sync is terminated on the desktop; otherwise, the connection between Palm and debugger cannot be established. Now follow these steps:

1. In the debugger console, type **open 0 "Unsaved Preferences."** This command opens the Unsaved Preferences database, which contains a scrambled password. You should get an answer like *Success! Access Pointer= 0000292A.* If a connection error occurs, the Palm debugger will report something like *### Error $00000404 occurred.*
2. Now we will find the resource record within the Unsaved Preferences database with the scrambled data password. Type the command **getresource -t psys -id 1.** The Palm debugger will search a specified resource record and display it: *Found at 800402DA: h~*"·:FH_ _ _ _ _V‹_UE#_j_qC_·*
3. Copy the data string into the clipboard and paste it into the notepad. Save the file as password.bin. Now this file contains scrambled password binary data.
4. To recover the original password, we should use the PCrypt utility created by @stake. This utility command line is:

   PalmCrypt.exe -[e | d] «less»ASCII | password block«gtr»

   We need to run it as

   PalmCrypt -d
   binary_encoded_password_string_without_spaces.

   In our test binary, the encoded password was

   68 05 7E 2A 93 B7 3A A4 46 48 BC FD E5 F6 DF 56 05
   8B 8A 55 45 23 9D 6A 02 16 EB 1A 71 43 80 B7

and we should run PalmCrypt as:

PalmCrypt -d
68057E2A93B73AA44648BCFDE5F6DF56058B8A55452
39D6A0216EB1A714380B7

It produces the following output:

```
Palm OS Password Codec
kingpin@atstake.com
@stake Research Labs
http://www.atstake.com/research
August 2000
0x74 0x65 0x73 0x74 0x2D 0x70 0x61 0x73   [test-pas]
0x73 0x77 0x6F 0x72 0x64                   [sword   ]
```

The original password was a test-password. Now we can remove the Palm from the cradle, reset it, and enter this password on the lockout screen. The device is then unlocked and unprotected.

The Soft and Hard Reset

Almost all PDAs support two special operations: soft and hard resets. A soft reset is similar to a reset on the desktop computer: The CPU is reset and the OS is booted and started. It finds the existing file system in RAM or ROM and mounts it. Data is usually not lost after a soft reset. During such an operation, the PDA may load and execute certain applications according to saved settings.

A soft reset should be performed only if the user experiences problems with the PDA software. For example, some applications may stop responding or use many system resources, causing the system to malfunction. Each device also has its own procedure for performing a soft reset. Although you can check the users' manual for details, the usual method is poking a pointed object such as a paperclip into a little hole on the back or side of the device.

Sometimes a soft reset may not solve the problem. For example, if the OS tries to load and execute corrupted applications during startup, it may cause a persistent error. In this case, the user should perform a hard reset. A hard reset is similar to a cold boot on desktop computers. The OS is restarted and it performs self-tests and some calibration routines, usually touch-screen calibration. The file system and other objects stored in the RAM are also recreated from the factory default settings. Existing data before the hard reset is lost. Because user data is stored in RAM, unlike a desktop, generally after a hard reset your device will be back at the factory defaults, losing any applications or data.

From the security point of view, soft and hard resets may constitute security flaws because a malicious user may be attempting to bypass a security application by using the hard or soft reset. However, the security application can defeat this attempt if it is activated by a system reset at startup. In this case, the security application restricts access to device resources effectively.

Palm OS and Windows CE devices act similarly for soft and hard resets. Usually, a soft reset may not disable any active security, but a hard reset cleans the device and makes it accessible to an attack. Some third-party tools, however, enable Palm OS-based devices with Flash ROM to remain secure after a hard reset.

The RIM OS-based device acts differently. It offers soft reset functionality, but no hard reset capability exists. Almost all the device memory is in Flash RAM, and only a small amount of cached data is stored in *static RAM* (SRAM). Thus, the user cannot perform a hard reset because Flash RAM is an independent power source and the RIM device doesn't offer a way to perform the hard reset. However, certain utilities enable Flash RAM to be emptied and the the initial (factory) RIM OS image to be uploaded.

GRADUATION

WHERE DO WE GO FROM HERE?

CHAPTER NINE

HANDHELDS

THE FIRST DECADE OF
THE 21ST CENTURY

PROCESSOR, MEMORY, AND
APPLICATIONS: FEELING THE POWER

Throughout the era of the modern computer, as each platform achieved a cost-effective critical level of power to support robust software applications, those changes affected the way the corporate world did business. The evolution of the mainframe from an expensive platform that required whole buildings, to being able to be deployed for a reasonable cost within a normal room accelerated the penetration of mainframes in corporations. Once the hardware became cost-effective to purchase and maintain, a broad range of software applications were developed to take advantage of the platform's powerful processing capabilities. Similarly, *personal computers* (PCs) needed a combination of power, speed, and supporting software applications to move from a niche business tool to a tool that broadly affects every facet of an organization.

Personal digital assistants (PDAs) have already moved from consumer roots to broad but shallow deployment within the corporate environment. In 2002, more than 74 percent of corporations were buying handheld devices directly.[1] With the continued introduction of faster processors, more memory, and

[1]Palm Analysts. *Mobility Magazine*, Issue 1, November-December 2002.

larger storage, the platform now has the ability to support robust data and applications. Global Enterprises are finding that hand-held devices allow them to react more quickly than ever to changes in their environment and to capitalize on opportunities. With critical information available to employees 24 hours a day, 7 days a week, decisions are made more efficiently and more effectively. This ranges from customer lists to technical specs to shared calendars.

ALL WE NEED IS BANDWIDTH — AND HERE IT COMES

Handheld devices have the storage required to support robust applications and data locally, the processor speed to manipulate this information, and a growing library of off-the-shelf software applications to facilitate this activity. Just as access to *local area networks* (LANs) and *wide area networks* (WANs) allowed PCs to replace the mainframe "green screen" terminals through gaining access to broad corporate data, wireless access to PDAs will extend similar capabilities to the handheld platform.

There is a tremendous amount of capital investment being made in the wireless industry. This investment not only recognizes the expected broad penetration of robust wireless networks into the corporate infrastructure, but it insures the success of the development of these networks. The wireless network technology will enable corporate and campus networks, high-speed cellular and wireless data networks, and community and public wireless hot spots. With wireless data availability achieving an ever-widening coverage area, it is not surprising that handheld devices are adapting to take advantage of these networks.

Providing powerful mobile devices with deep access to corporate data through wireless networks can allow for more timely and effective decision-making by an increasingly mobile work-force. In the early adoption stage of these technologies during the first decade of the 21st century, those organizations that understand how to take advantage of this powerful technology have an opportunity to create distinctive competitive advantages. The deployment of mobile technology alone will not create value; the technology must be aligned with business architecture, strategy, and objectives.

THE COMMUNICATOR: DEVICE CONVERGENCE

Handheld device manufacturers recognize the importance of linking a device with the broader world around it. The power of a PDA is limited unless it can seamlessly access data from a variety of sources. Historically, PDAs used hotsyncing and infrared to access data, either sending information to a PC, or receiving data from a PC. New approaches that are becoming mainstream, such as Ethernet-based sync-servers, wireless networks (802.11b protocol), and Bluetooth, allow significant amounts of data to be delivered to handheld computing devices in a diverse environment.

In addition, a growing number of devices merge cell phones with PDAs. The power of these converged devices is that they allow the PDAs to use the proven, robust, and rapidly improving cellular networks to transmit data. For the user, converged devices help eliminate device clutter that plagues the early-adopters of mobile technology. These people carry cell phones and PDAs, as well as pagers and back-up cell phones, creating a mix of devices that become impractical to use.

The PDA/cell phone hybrids come in a variety of choices. These reflect the inherent compromises users must make between the processing power, screen size and quality, and applications available on PDAs, and the form factor and familiar interface of cell phones. High-speed data transmission, integrated text messaging, web browsing, on-device games, and high-resolution color screens have already become standard on mid-range cell phones. Integrated digital cameras, basic email capabilities, and integration with desktop calendar and address-book programs are also becoming increasingly standard on higher-end models. Cell phones already link laptop computers to the Internet, taking advantage of connecting comparatively "giant" laptops to corporate networks, email, and the Internet. Handheld devices, connected through Bluetooth, cable, or modem card, or directly integrated with cell phone technology, bring this same connectivity to a much more portable and accessible platform.

Merged devices, which bring a fully functioning PDA with embedded connectivity through access to the cellular networks, will come to dominate the upper-end of the mobile device market. However, Smartphones, or cell phones that integrate to some extent with desktop data, are becoming mainstream. Given the tremendously higher number of cell phones as compared to the full-featured PDAs, these "smart" devices will play a key role in the drive toward bringing data mobile. Although many players participate in this market, including the leading PDA software companies, it appears to be a two-horse race to dominate the Smartphone space.

Microsoft has made a clear commitment to adapt its WinCE/PocketPC software to the Smartphone market. Microsoft recognizes that the future of these smart devices relies on seamless connectivity to information that resides on the desktop, be it calendars, address books, or broader corporate databases. Microsoft's desktop dominance allows for robust connectivity between the mobile device and the desktop data, and can do so leveraging a "user interface" that is familiar to most PC users. In addition, Microsoft has the capital and patience to invest in this market.

However, the devices supporting the Symbian *operating system* (OS) may prove to be too formidable a competitor for Microsoft to monopolize the Smartphone market. Symbian was established as a private independent company in June 1998 and is owned by Ericsson, Nokia, Matsushita (Panasonic), Motorola, Psion, Siemens, and Sony Ericsson. Symbian has created a robust OS focused only on the mobile market. It is attempting to create a standard set of *application programming interfaces* (APIs) within a robust, yet compact OS that specifically enables integration of data into cell phones. With so many major cell phone manufacturers invested in Symbian, already a leader in Europe and Asia, it seems certain that Symbian will dominate the Smartphone market soon. In the long-term, however, the strength of the OS and its ability to meet the needs of the Enterprise and consumer markets will determine the winner of this battle. Given Microsoft's capital and success in so many other markets, it should never be counted out. The market opportunity is so significant that we expect the company to continue to make the investment necessary to innovate and capture market share.

The question isn't whether to buy a PDA that has direct access to cellular networks, but what compromises to make in features, cost, form factor, and performance. On one end, we are seeing the emergence of traditional PDA form-factored devices that are adding cell phone functionality. The BlackBerry recently added cell coverage to its existing form factor. However, it maintains the form factor and functionality that have made the BlackBerry the leading mobile email niche device on the market today. Likewise, early implementations of Smartphones based on PocketPC 2002 platform choose to maintain the traditional form factor of the PocketPC at the expense of seamless phone convenience and compact form factor.

Kyocera and Handspring continue innovative convergence of cell phone and PDA features. Each has taken a unique approach to make sure that the cell phone features (ease of use and form factor) are not subsumed to PDA functionality. The June 2003 announcement of Palm, Inc.'s prospective acquisition of merger with Handspring may create another credible competitor in this market. However, this is dependent on the combined company's ability to capitalize on Palm's name recognition and stable operating system. The new Kyocera 7135 embeds the graffiti/data-entry area into the folded interior of the cell phone to allow reasonable screen size and traditional data entry with a small and ergonomic form factor. Innovation in key areas will enable Smartphones to become the predominant handheld computing device. Innovative form factors will increase the computing capabilities in traditional cell phones.

The bandwidth will not only increase, but will become more affordable as more high-speed networks come into service and compete for customers. Finally, advances in software will not only enable seamless transition among phone, data, and computing tasks, but will integrate them to bring timely and relevant information to the palm of your hand.

IT'S NOT WHAT, BUT HOW AND WHY

So handhelds are the next hot platform. The power of desktop PCs is accessible from a handheld device, with high-speed connectivity to broad networks. Like many innovations, handhelds will not replace what has come before, but will become a new component within the corporate infrastructure. The question is not when and how deeply these devices will come into play, for they are already here. Now we face the task of how to maximize the corporation's return on this technical infrastructure investment. In this new century, corporations are focusing on understanding technology's *return on investment*, and the increased penetration of handheld devices into the corporate world depends on how effectively hardware vendors, software vendors, and integrators understand how to translate the technology into financial returns.

With their high, inherent failure risk, mobile applications are no different from other technologies inserted within an organization. Leading organizations minimize this risk through tightly aligning the technology with their business factors. Understanding how to use the mobility of information to allow the organization to make decisions faster and meet the needs of its customers, suppliers, and other constituents is not an easy matter. The various aspects of mobile devices dictate a heterogeneous implementation of the technology, with each organization needing to incorporate and support a range of devices depending on its unique requirements. "One device fits all" does not work just yet, if it ever will. Handhelds can support anything from simple *personal information manager* (PIM) functionality to complex disaster response scenarios in which real-time video, extensive data access, and location-based information tied to *global positioning system* (GPS) technology can make the difference between life and death.

CORPORATE MARKETS WILL
DRIVE INNOVATION

The difference between success and failure in the competitive corporate marketplace is based on simple, yet fundamental factors. Timely, accurate information is critical to effective decision making. The mobility of information between the corporation and its stakeholders (customers, employees, vendors, and others) allows for better and faster decision making.

To date, we have seen rapid innovation with respect to the physical capabilities of the mobile device hardware, combined with enhancements in the key OSs to support the increased connectivity, processing power, and storage. The next stage of real innovation will move from hardware and OSs to applications.

Corporations recognize that handheld devices are becoming increasingly important in corporate information infrastructure. They are investing money in hardware, developing and purchasing software, and integrating these devices into their businesses. The impact of this investment will be the mainstream extension of streaming data to these devices. Video will be able to be communicated from the world at large and transferred back to corporate leaders. Devices with access to broad networks will be able to capture sites, sounds, and images from the edge of the corporate boundaries and transfer that back to the decision makers. We can already imagine emergency responders to disasters transmitting ground zero images back to teams to analyze and enable effective response. This ability to send mobile devices to gather data out in the world, transmit it back to people and systems that can analyze such data, and then return actionable information back to the point of contact will significantly change a broad range of industries and organizations.

LOCATION, LOCATION, LOCATION: THE POWER OF GPS

Handheld technology is capable of significantly affecting the world, just as reliable DNA testing allows us to determine facts that were heretofore a mystery, from questions of paternity to the guilt or innocence of alleged criminals. Information captured on and transmitted by handheld devices goes beyond just what a person can perceive, all the way to chemical sensors and analyzers and location-based information driven by GPS technology. Seamlessly connecting data not only to time, but also to place and other environmental activities, brings a range of information that will significantly enhance the usefulness of this information. Handheld devices can capture video, still images, sounds, environmental conditions (such as chemical readings, temperature, and pressure), location, and other data, and seamlessly transmit it via high-speed wireless networks. This information can then be processed within an organization, and possibly against other data being simultaneously collected. Actionable data can be immediately transmitted back to the field, telling the fireman to get out of the building that's about to explode, the shareholder to sell the stock that is starting to drop, and the salesperson to mark down the inventory on the sales floor.

Integration of GPS into mobile devices lets people know where they are, what is relevant around them, and how to get where they need to go. It also can be used to track and monitor people, animals, and objects. The truck carrying valuable cargo can have its progress perpetually monitored. Systems can alert decision makers to problems, customers to real-time delivery status, and emergency responders to locate the required resources when they need them.

BUT LET'S NOT FORGET THE CONSUMERS

Although the corporate market is driving the broad handheld device market, consumers are clearly driving the cell phone

markets. With significant penetration and growing capacity, the cell phone industry is looking for ways to increase its revenues and profits from the investments made in networks and infrastructure. Data transmission was thought to be a key revenue driver, with *short message system* (SMS) messaging becoming widely accepted, especially in Europe and Asia. However, it is clear that games (single and multi-player games), transmission of digital images captured on cell phones, and other software applications are becoming key drivers for the cellular companies in the United States as they attempt to profit from the proliferation of phones and the availability of bandwidth.

As the distinction between phones and PDAs continues to blur, this large consumer market will influence the hybrid devices. Leading manufacturers such as Sony are pushing PDAs and phones toward the entertainment markets, integrating the devices into their mainstream consumer product lines with links between cell phones, digital cameras, video games, and desktop computers. As bandwidth to these devices expands, carriers will profit by bringing streaming audio and video to phones and PDAs. Music, movie clips, and other information will be accessible around the clock and around the world. The location-based capabilities of GPS enable new ways of interacting with our environment. Driving through town, we can find out which is the nearest movie theatre, assess movie schedules, view preview clips, and order tickets—all from a PDA. The technology is available to do this. It awaits only bandwidth.

The ramifications of such technology goes beyond the user's experience. Are we to allow marketing companies to track the movies we watch? To track where we are physically located when we decided to go to the movie?

All this data will be captured in the telecommunications companies' systems and will be available for companies to analyze and use. This data can allow the corporation to better understand its customers and improve its services. However, they can also use this information to change behavior, sell information, or leave it exposed to people with malicious intent. As on the Internet, a battle for privacy rights will drive a need for well-stated and strongly enforced privacy policies.

WHO WILL WIN THE OS PLATFORM WARS?

As has been discussed in this book, a variety of OSs power today's mobile devices. The early lead, with respect to number of deployed units, is held by Palm, whose handheld devices made the term "Palm Pilot" ubiquitous. Palm's OS is marked by efficiency, strong developer support, and a huge amount of legacy applications that enable Palm Pilots to be of use to a broad segment of the market.

However, Microsoft's mobile OSs based on WinCE, including PocketPC and PocketPC 2002, clearly are making an impact on the market. Microsoft's technology seems focused on a ".Net" implementation, allowing software code and devices to be portable from handhelds to desktop devices. Microsoft's strategy to implement handheld devices into the broader *information technology* (IT) infrastructure not only is powerful in and of itself, but also leverages Microsoft's unique monopoly on the desktop OS market. Being able not only to link the platforms through software, but to control the code underlying both platforms puts Microsoft in a unique position to execute this strategy.

Linux, to a lesser extent, has the same opportunities, as it picks up segments of the market that have interest and expertise in the Linux desktop and servers.

Symbian is dominant in Europe and Asia, and has many of the same attributes going for it as Palm OS. It is flexible, light, efficient, and designed specifically to support mobile devices. While Symbian doesn't have Palm's global market share to leverage, its supporters (and shareholders) are some of the leading global cell phone manufacturers, thereby insuring its spot as maybe the strongest player in the Smartphone market.

Who will be the winner? No one will dominate to the extent that Microsoft dominates the desktop. Symbian will maintain a position in the global marketplace, as long as it can keep the support of its hardware vendors. Likewise, Linux will continue to have a place in scientific, education, and business applications in which Linux-skilled developers might have an interest.

Palm will maintain a very strong position in the lower- and middle-range handheld devices, focusing on its core strengths of low hardware requirements (lower-costing devices) and a large installed corporate base. Whether Palm has the resources and creativity to grab the upper end of the market back from Microsoft OS devices remains to be seen. In the end, however, none of the other vendors or partners supporting the competing OSs have the capital, the marketing talent, or the monopoly on the desktop to truly compete with Microsoft on the upper-end devices. Handheld markets will see the same trends of pricing for processing power, storage, and *random access memory* (RAM) that affect the PC markets. With lower-priced, more powerful hardware, Microsoft's OS will be able to accommodate lower priced hardware and allow PocketPC devices to be available to the low- and mid-range markets. We can already see this in the recently released Dell devices. With a street price of $250, Dell has already brought down the pricing on other entry-level PocketPC devices.

Can Palm compete? We think so. Palm's new devices compare favorably with the PocketPC. If Palm OS 5.0 (and 6.0) delivers on its promise of supporting more robust applications, more powerful processors, and more diverse features (native MP3 players), Palm has a window to succeed. If Palm has not recaptured the market's interest by the end of 2003, it will have difficulty ever effectively competing with Microsoft, Symbian, and other niche handheld OSs.

MASS CUSTOMIZATION OR A COMMON DEVICE METAPHOR

The emergence of PDAs as a mass market device was enabled by the broad success of a single vendor. Other vendors had marketed handheld computers before the emergence of the Palm Pilot, but Palm's mix of function, features, and price became widely recognized and accepted, despite an interface that was not particularly intuitive. Palm's successful implementation of technology enabled the era of the handheld computer. Because of Palm's dominant market share, its form factor and core functionality became the common device prototype for the early stages of the handheld market. Early PocketPC devices resembled Palm Pilots as far as the method of data entry, core functions, features, and form factor. Even the PocketPC maintained the same device design, but added support for faster processors, integration with Microsoft software, and a new interface that resembled the familiar Windows.

As the market emerges, so does form factor innovations. The high resolution and flipable screen of the Sony Clié, combined with an integrated digital camera, enables the device to display information as well as access it. Integration with other Sony consumer products via the Sony memory stick allows the PDA to become the interface for a large number of devices and applications.

The merged devices also create significant variations in the form factor. Using a PDA to provide driving directions on a GPS-enabled map will most likely require a significantly different form, screen, power, and applications than the device that maintains the calendar for the "soccer mom." Likewise, the first responder in a domestic disaster will require far different functionality than other people, not only in method of accessing and communicating information, but also in the "ruggedness" of the device.

As manufacturers wrestle with the cost of accommodating users, we are already seeing device divergence.

The key factors driving the growing number of different device designs are not the core features, but the peripherals attached to and embedded in the devices. PDAs are being connected to cameras, GPS devices, MP3 players, bar code devices, chemical sensors, and streaming video capture hardware, among other devices. The handheld computer is a platform that gathers, stores, provides access to, and displays data. Only the manufacturers' imagination and the customers' pocketbooks will limit the diversity of uses and configurations. Fortunately, neither of these limiting factors appear to have any impact on the continuing innovations in this field.

The primary uses of the devices have a significant impact on the implementation. Does a customer want a Smartphone, a PDA capable of making a call, or an email or paging device that can open attachments? A common device having similar form factors, features, specifications, and pricing will not emerge as it has done in the PC market, but instead we will see varieties of device functionality based on the principal market. No clear lines will exist between these functions, as the devices will maintain blended transitions based on shared and predominant device functionalities.

The Smartphone might contain applications (games), data (phone numbers/addresses, calendars) and Internet access, but it will still look, feel, and act like a phone. Key features will be size, form factor, and inclusion of all key features expected on a cell phone. However, these devices will lack large screens, will be difficult for data access/entry, and be of limited quality (to keep the retail price down). Some manufacturers are pursuing higher-end "merged devices" that offer PDA features within a unique implementation of phone features. The Kyocera model 7135 mentioned above has a larger screen and will include an integrated data-entry area and high-quality color screen. However, it is larger than the mainstream cell phones and also has a smaller screen than full-featured PDAs. The Kyocera is a good example of the compromises that users will make as they determine what they want from their mobile devices.

Another innovator is Nokia with its Communicator. Nokia produced a full-featured, traditional cellular phone that opens up to reveal a QUERTY keyboard, a high-quality, color screen, and seamless Internet capabilities. It integrates the phone and PDA features well, but it is still too large to be a primary phone. Future versions will surely continue to improve the form factor, the storage, processor, and memory for this device. Many other good examples of innovative merged devices can be found as well. Expect to see the continued integration of phones with PDAs, while manufacturers continue to make compromises, depending on the primary use and environment of the PDA. The traditional PDA will continue to maintain the Palm Pilot device design. Graffiti (or another, similar Stylus-based entry), syncing of data with networks and desktop computers, simple, intuitive interface, long battery life, and a host of applications will dominate this segment. The legacy of the Palm Pilot interface will be that these lower-priced, functional devices will maintain that familiar feel and price point, with increasing power and functionality.

Niche devices will emerge that implement industry- and application-specific form factors to support specific markets. An example is the Symbol devices that run on the Palm OS and directly integrate scanning capabilities into the hardware. The form factor of the device and the OS (or at least the application interface) will focus on bar code scanning to support mass mobile data entry. Wireless integration, security, and higher quality components will affect this area to support the requirements for connectivity and data entry in fields such as manufacturing, health care, and retail.

Expect to see continued niche implementation of PDAs with form factors becoming more diverse as people try to bring connectivity and information out into the world. Expect mobile devices that have little user interface, but are deployed to gather, communicate, and access data. Integration of mobile computing into clothing has already been demonstrated and is being considered for a variety of applications. Imagine the soldier going to battle wearing his or her PDA, with access to data through voice commands, or a flip-down ocular. This PDA will automatically collect environmental data without the soldier's involvement and wirelessly transmit it back to systems and individuals. More important, the data from the battalion will be transmitted wirelessly and correlated seamlessly with vast amounts of other information to enable better decision making. Such decisions will be transmitted back to the battalion, with the PDA becoming the link between the soldier and the command.

The tablet computer is another example of using a form factor and an interface to change the nature of the handheld device. Now the tablet computer can be used on the shop floor, integrated with work instructions, technical manuals, and manufacturing systems. Wirelessly connected, these devices can reduce manufacturing costs and increase quality by bringing the network to the worker.

It is not likely that geography will be a significant, long-term factor in the device design. Today, geographic distinctions relate to devices and networks. Symbian OS and Nokia devices have a much greater impact on the European and Asian markets than they do in the United States. However, as PocketPC and Palm make inroads into Europe and Asia, and Nokia introduces its Communicator in the United States, the global market grows. This market has differences with respect to language and OS, but this does not seem to be a long-term trend. In fact, the major vendors all recognize that the handheld market is global and that they need to take a global strategy in approaching it.

WHERE DOES THE DEVICE SECURITY GO FROM HERE?

The power, speed, storage, and applications available on mobile devices will significantly increase people's use of and reliance on their *personal digital assistants* (PDAs). They will have their calendar, task lists, email, and a vast array of personal information on these devices. This window to each person's world will be connected to the World Wide Web, will have Bluetooth ports open, can be called, and can be physically stolen. The frightening combination of personal information on mobile devices, theft, and the easy connectivity by would-be hackers mandates significant enhancements to security on PDAs.

With current news being awash with credit card theft, identity theft, and recreational mayhem by young hackers looking to show off their talents, the risk of having so much personal information residing on a mobile, connected device is significant. These issues are exacerbated when the information stored on a device is extended to other people's information: Company information, patient information, and customer information are all increasingly found on PDAs.

In analyzing the early-adopter organizations that are proactively focusing on Enterprise-grade PDA security, two clear segments have emerged—the healthcare industry and government organizations, including the military. These segments have particular sensitivity to security issues and clearly highlight the key issues related to PDA security.

Doctors are increasingly using PDAs to bring patient information to a mobile setting, to access drug and medical information, to electronically transcribe patient information and for many other activities. A survey released in November 2002 by the *Healthcare Information and Management Systems Society* (HIMSS) stated that nearly 72 percent of physician offices have doctors who practice medicine with a handheld computer of some type, such as a PDA. In February 2003, the U.S. Department of Health and Human Services issued the final adoption rules for the *Health Insurance Portability and Accountability Act* (HIPAA). HIPAA applies to data stored electronically, which would encompass PDAs. Transmission standards in HIPAA apply when data is transmitted wirelessly. Noncompliance with HIPAA carries significant monetary penalties. Therefore, security and encryption are integral to maintaining compliance with these standards. More important, if sensitive individual patient information is stored on a PDA that has inherent security vulnerabilities, it is mandatory to provide sufficient security for patients to feel that their personal data is protected.

Even if some doctors are not yet focusing on these areas, the hospitals that support the doctors understand the potential liabilities and are addressing these issues. What is the impact on a patient whose history, diagnosis, and treatments are exposed? The potential impact is heightened by the ease with which information can be broadly disseminated over the Internet. Imagine a variety of websites springing up to capture, display, gossip about, and profit from health information. Extend this image to health or psychiatric information about celebrities and envision the proactive efforts by a variety of people to obtain confidential information. We can see that the PDA will continue to be one of the most vulnerable sources of such information.

Military and homeland security organizations are just as inherently sensitive to security issues. PDAs have been embraced by the military. On the one hand, PDAs are coming into these organizations just as they do in companies: People buy them with their own money and then sync them to the network, putting data, email, and contacts onto the device. In addition, the military has recognized the benefits of getting information into the hands of the war fighter, wherever they might be. The U.S. Department of Defense has embraced a network-centric computing platform, whereby data can be accessed anytime from anywhere through a standard browser, and it is developing methods to include access from mobile devices. In the Cold War era the world was divided, and the threats—while sophisticated—were at least known. Information was isolated within organizations, with safeguards over both electronic and written information. In tomorrow's world, information will be both portable and accessible, and the world will be electronically connected. That means that not only more information, but more sensitive information, will be stored on and accessible from mobile devices.

Examples of PDA use by the military are abundant. From Navy personnel accessing maintenance information on the flight deck of an aircraft carrier, to medical personnel wirelessly accessing information from the field, PDAs are clearly helping to accomplish military missions. While the military does have some unique issues, it faces the same security issues as other organizations.

INDIVIDUAL DEVICE-LEVEL SECURITY

The first level of security is the on-device security. The consumer market drove the evolution of the PDA, with many compromises made related to security. There is an inherent tradeoff between security and ease of use; the more comprehensive the security, the more difficult it is to easily access information. Throughout this book, we have discussed various mechanisms available to secure mobile devices. They include strong password protection, strong encryption of the data on the devices, bit-wiping data off the device, security over all access points (syncing ports, infrared, wireless, and so on), and protection for memory cards.

In looking at trends in security, we see that device-level protection will be enhanced and modified to accommodate the rapidly changing hardware platforms and options. A growing number of devices and hardware plug-ins will create unique security issues. Connecting a *global positioning system* (GPS) device to a PDA suddenly allows geographic information to be continuously sent and received from the device. It may be dangerous if someone can access your device and track your location from the GPS equipment. Likewise, the always-on connectivity related to PDAs with Bluetooth creates a constant security vulnerability.

The first line of defense is protecting information on the device. PDA hardware manufacturers are starting to upgrade embedded security by enhancing the *operating system* (OS) security, as well as licensing security applications from the *independent software vendors* (ISVs), and bundling this more advanced on-device security with more expensive product configurations.

The Palm OS 3.0 contained only very basic security. It could be protected by a password, but was not automatically locked when it was powered off. The number of keystrokes required to lock the device made the consistent use of the security features impractical. Private records were not encrypted. Information on the device, regardless of the user's utilization of the included security options, was accessible through commonly known methods.

By contrast, Palm OS 5.0 includes more extensive security features. The user may automatically lock the device on powering off, and some of the commonly known backdoors into the device have been closed.

While security on PDAs has improved, hackers' skills and sophisticated software that bypasses security features renders embedded security inadequate. Even more inadequate is security on Smartphones and merged devices. Smartphones come with memory, basic *personal information manager* (PIM) functionality (such as calendars and phone directories that sync with information on the user's desktop), Internet browsing capabilities, and other features that enable the cell phone to function much like the early PDAs. Traditional security for cell phones focused on preventing their unauthorized use rather than protecting information on the devices. We will soon experience more demand for enhanced security on Smartphones. Security vendors now focused on PDAs will move to protect all mobile devices, from phones to traditional PDAs.

Finally, merged devices—which combine fully functioning PDAs with cell phones—create unique security issues, due to the need to access certain data immediately, without security, while other information needs to be very secure. You don't want a lot of security restrictions when you are trying to answer a phone call, but credit card information should be very secure. In the emerging cross-functional device market, each manufacturer has merged the phone and PDA features uniquely, some in the hardware, others through modifications to the OSs and additional software. Since robust security for PDAs is primarily provided by third-party mobile security software companies, they will need an adjustment period to respond to the manufacturers' implementations. In the meantime, security-minded buyers will have to consider both the features of a device and the security software available for it. The hardware vendors that grab an early and significant market share for these merged devices will provide incentives for security vendors to accommodate their platforms. Devices without sufficient market penetration will most likely either be competing in a market without robust security, or will have to invest in security themselves.

In addition to embedded enhanced security, we are seeing an emerging interest in including biometric security on the PDAs. Biometric security includes retinal scans, fingerprint recognition, electronic signature recognition, and even voice recognition. The newer devices are capable of processing biometric security features.

The ability to connect other hardware to the new PDAs greatly expands their security features. Capturing electronic images through a connected, high-resolution digital camera allows for a cost-effective retinal scan or other similar biometric access capability. Even cell phones are now being marketed with embedded digital cameras. Ultra-security-minded people might have to scan their retinas before making a call on their cell phone or accessing the private address book.

Another innovative and easily deployed biometric security feature now available is the electronic signature, which is marketed by CIC (www.cic.com) and Cloakware (www.cloakware.com), among others. Device access is restricted based upon a biometrically validated signature which the manufacturers of such products claim to be unique to individuals. As the technologies mature, biometric security brings enhanced, personalized access control over the devices.

As with desktop security, mobile embedded security will become a combination of niche products such as device security, firewall, encryption, and anti-virus bundles. A few market leaders will emerge in each of these key areas, with some vendors moving toward comprehensive products that address all. Other vendors will focus on specific niches and enhance the ability of the niche products to work with other market leaders. For example, a user might match anti-virus software from Symantec with on-device security software from PDA Defense and Certicom's wireless encryption products to encrypt data over the Internet. These combined products provide a comprehensive range of security protection. Increasingly, hardware vendors will bundle trial security products with the devices, and security-sensitive users will use and upgrade these products.

Security, encryption, and anti-virus products are becoming standard for individual PC users, even home and small office users. As people are educated about the risks related to mobile devices, the use of security and anti-virus products will increase. The likelihood that confidential information will be stored on a PDA is high. With the number of phones and PDAs that are lost and stolen each year, public awareness of the issue will only increase.

SECURITY WILL CHALLENGE THE CORPORATIONS

Unlike individuals, who understand the issues of mobile security but aren't always motivated to address them, Enterprises are already doing so. The risks to a corporation from lost or stolen devices are enormous.

Today, both large and small corporations are trying to address mobile security issues at the policy level. While developing policy might seem easy at face value, the reality is that it is often difficult to enforce security policy over devices that are owned by employees and not by the company. Once these devices touch the corporate network, they have access to core corporate data. Large memory capacity, integration with memory cards and other devices that can store additional data, and access to wireless networks make it impossible for corporations to ignore the exposure inherent in these devices.

Once the mix of corporate and personal devices is acknowledged, corporations must deal with issues surrounding deployment of PDAs within the organization. These include help desk support, budgeting for the purchase of the physical devices, security surrounding the products, and most important, security software, policies and enforcement.

As we discussed earlier, software is already available to define security policies and deploy them onto the device. We also described software that allows *information technology* (IT) organizations to manage the devices, whether or not they are connected to the company's networks. Overall, however, the PDA security industry is still immature. Improvements are needed to address hardware platforms, OS platforms, and software applications, both off-the-shelf and custom. Existing products will mature, providing cross-platform, fully featured security that is centrally managed and enforced.

Standards will slowly emerge that address the interaction between security, hardware, software, and wireless networks. What this means for software developers is that they can create software that understands how to work with the leading security products. Encryption will be accessed through standard *application programmer interfaces* (APIs) that will allow for seamless encryption and decryption of data managed by the OSs and/or software applications. This standardization is necessary for hardware manufacturers to create new devices without creating conflicts with the leading security products on the market.

Today, corporations generally must limit themselves to a few types of devices because no standard security products are effective for all devices and all configurations. As new converged devices come on the market, mixing traditional PDA function-ality with phone, MP3, GPS, and other features, the gap between availability of the devices and the ability to secure their data will disappear. When that happens, corporate security can be enforced on a range of devices with a standard security prod-uct or suite of products.

PDAs Become the Access Point to Corporate Networks

While walking through a neighborhood department store, I noticed a Palm Pilot embedded in the wall, with only its screen showing. Upon closer examination, I noticed that the Palm Pilot had been integrated into the store's alarm system and was used to provide data entry and information on both the fire and alarm systems. This demonstrates not only the flexibility of the PDA interface, but also Palm OS's capability to govern access to other systems.

PDAs are able to authenticate users to the device and then use that information to govern access to other systems. Imagine capturing biometric information such as a validated thumb print and then using that information to obtain access to files over the Internet, to broader corporate intranets, and to physi-cal locations. If this biometric information is in your cell phone, perhaps validated through a retinal scan taken by the embedded digital camera, you then could transact business from your phone by accessing credit card accounts, getting cash, or maybe even making a phone call.

Now, corporations can have small, mobile devices that can be accessed only by authorized employees. Access is controlled by as little as a password, and at most by a combination of a password, a physical device (Bluetooth or plug-in physical authentication device), and/or a biometric verification. Once someone has authenticated against the device, they now can have full "role-based" access to corporate data, networks, and facilities.

Mobile devices are being equipped with wireless modems and Bluetooth capabilities, and continue to have the traditional access points of Infrared and docking via serial or *universal serial bus* (USB) ports. All these avenues can be used to access corporate networks. Mobile *virtual private networks* (VPNs) will provide the same secure wireless and dial-up connectivity that is currently available on remote PCs and laptops. A number of products now on the market encrypt data being sent wirelessly from a PDA to a network by establishing a VPN, as illustrated in Chapter 3. In addition, many organizations are relying on secure socket layer (SSL), which supports secure transactions over wireless connections, and which has become standard for browsers on PDAs and cell phones.

Over the next few years, a few products will lead the way in enforcing mobile-device security policies. They will track devices that attach to the networks and log such information as which applications are running, when the devices touch the corporate network, and security events and breaches that occur on the devices. With the wireless capabilities of the new generation of devices, each event will be transmitted and logged as it occurs; intelligent monitoring and notification systems will insure that the corporation is able to respond quickly to breaches. These features will apply against all platforms and include the Smartphones.

As security capabilities become more sophisticated, and connectivity between remote devices and corporate networks becomes "always on" and real time, privacy issues will become even more important. Although the United States has seen delays in implementing the use of cell phones with GPS for 911 emergency responses, the technology to locate every cell phone will be implemented within the next 10 years. Being able to track cell phones to a specific location will save lives and allow police, firefighters, and ambulances to respond more quickly and effectively when emergencies are reported on a cell phone.

With homeland security a concern around the world, cell phones with GPS location tracking can help coordinate disaster responses. If a robust system can track the physical location of all medical workers, ambulances, firefighters, and police in real time, these resources can be coordinated effectively. A doctor can be directed to an area hospital based on the doctor's location rather than his or her home hospital. In addition, emergency responders can communicate rapidly via their GPS phones and PDAs by using automated voice response systems such as that marketed by SurfSimple (www.surfsimple.com). This technology enables a system or person to quickly send a text message to any number of cell phones with the text message converted to a voice message.

Carefully Assess and Document Privacy Policies Related to PDAs

The power of GPS, of networks that can track activity back to each individual device, and of software that can monitor all applications on each device, give the corporation access to potentially sensitive information related to an individual. These smart devices will know who you are, where you are, and when you're there. It's a little frightening. What kind of protection will individuals have with respect to their privacy? Can a corporation fire someone based on knowing where they are and what they are doing? Does a corporation have to explicitly request permission to monitor such activities, or does providing a corporate Smartphone with GPS give the company the right to monitor such information?

The use of Smartphones and PDAs for both business and personal purposes only complicates this issue. If the corporation must provide security over all PDAs that touch its networks —even those that individuals have brought into the corporation —does that give the company the right to monitor all activities? And what rights does the corporation have to personal information residing on a corporate-owned device?

Corporations must proactively assess privacy issues related to handheld devices and Smartphones. The amount of information available to be logged, analyzed, and retained related to information on the devices is huge. Corporations risk litigation if information is retained and used without consent, or if the information is not retained and therefore not available to support certain decisions. The existence of such information could also adversely impact a company in the case of litigation. The availability of such a wealth of information in the hands of skilled investigators and attorneys could prove to have adverse consequences in certain litigation situations. The implications of having access to and using such information must be carefully addressed at a policy level, and that policy must be documented to support the decision, and should be communicated to company personnel.

Likewise, companies that provide services related to PDAs also need to be careful with respect to privacy issues. Tracking people's locations can be useful for marketing purposes. Location-based advertising can be sent to PDAs to advertise items and events physically near the target. Likewise, this kind of marketing can be a huge invasion of personal privacy. Phone companies, corporations, and other companies should carefully develop and declare privacy policies proactively so that individuals feel confident accessing the fantastic benefits of location-based technology. Our legislators must proactively assess legal restrictions on the use and admissibility of such data.

Authentication, Authorization, and Verification

The story with respect to PDA security is really the same as all other computer platforms. PDAs have evolved more quickly than any other computing platform, and the growth in processors, memory, functionality, and market penetration will continue to accelerate rapidly.

Security software will continue to evolve, to cover all key platforms and devices. The lines continue to blur between the laptop, the PDA, and the cell phone. Enterprise security will have to address all mobile devices in a consistent and proactive manner.

As with computers, the key to effective security is: To authenticate—to make sure only valid users can access the device; to authorize—to make sure that only the authenticated person gets access to the information to which they have been granted rights; and to verify—to provide tracking, logging, and auditing so that security policies are enforced.

Mobile computing devices are the next phase in the information age. They bring the information from your desktop PC, from the Internet, and from the corporate network to wherever you might be. Having access to information anytime and anywhere, along with connectivity to the outside world, will allow people to stay connected in ways that have never been possible. However, along with connectivity and mobility comes the requirement for security. Proactive security, starting with good corporate policy and implemented with effective Enterprise mobile security software products, will enable individuals and organizations to realize the tremendous productivity and personally enriching benefits of having the world in the palm of your hand.

INDEX

I

Q–R

ABOUT THE
AUTHORS

David Mclnick currently holds the position of President of PDA Defense, the leading Enterprise PDA Security offering. Prior to PDA Defense, David helped to pioneer Internet- based financial transactional systems and security from the mid- 90s publishing multiple books on the topic and consulting with a wide range of companies on the execution of electronic commerce systems. As early as 1996, David was the focus of case studies in the area of Internet-based Electronic Commerce (Microsoft's Merchant Server Book—Ventana, 1996). And, by 1997 David had taken the lead author role in the first book published on Microsoft's Active Server Pages (QUE/Macmillan Publishing, 1997).

From the beginning of the 1990s David has worked to leverage technology and manage risk within the Financial Services industry starting with his work for one of the largest credit card processors. At GE's Credit Card companies, David worked with GE Corporate R&D to implement the first commercial application of Neural Network technology for risk management scoring within the company, as well as later operationally working in both Fraud management and Risk Scoring Development groups. After GE, David's financial services work continued with the architecting of payment gateways currently in use by Bank of America and Wells Fargo Bank. In the credit card processing area, David worked closely with Visa and Bank of America to pilot early tests with Smart Cards and emerging security standards such as SET 0.0 for taking transactions over the Internet.

Through a combination of published work, speaking engage-
ments, and professional organization involvement, David has established a reputation for leadership in the technology-intensive security
and risk management areas, including a specialization in the area of
financial services.

Mark Dinman has over 17 years of experience in software systems
development and IT project/program management. Mark joined
Asynchrony Software when it launched in 2000 and served as the
product manager of PDA Defense since its inception. Prior to joining
Asynchrony, Mark spent 15 years with EDS in various roles serving a
wide variety of clients and industries. His last assignment with EDS
was as a Program Manager responsible for a global team developing
and supporting a Material Management System for General Motors.
Mark earned his master's degree in business administration from
Washington University's John M. Olin School of Business and his BA
magna cum laude from Brown University. He is also a certified Project Management Professional by the Project Management Institute
(PMI) and currently serves as the President for the St. Louis PMI
Chapter. Mark lives in Chesterfield, Missouri, with his wife Michele
and his two children Jonathan and Bryan.

Bob Elfanbaum is the Co-Founder and President of Asynchrony
Solutions, Inc., a software technology firm with products and services focused on security, collaboration, and systems integration,
including PDA Defense Enterprise, a leading security solution for
handheld devices. Since its inception in 1998, Bob has led the company to significant relationships with Fortune 1000 and federal government clients including the U.S. Department of Defense, Boeing,
and McKesson Information Solutions. Prior to founding Asynchrony,
Bob was Chief Financial Officer of Virbac Corporation (NASDAQ:
VBAC), a public animal health company, managing both the company's systems and financial organization. Bob is a CPA who also
spent two years running the internal audit division of a Fortune 500
company, as well as eight years in the audit group of Price-Water-
houseCoopers. He earned his Bachelors degree, cum laude, with a
major in accounting from Indiana University, Bloomington.